GeoPlanet: Earth and Planetary Sciences

The GeoPlanet series is a forum for presenting the latest achievements in the Earth and space sciences. It is published by the GeoPlanet consortium (Earth and Planetary Research Centre) formed by five institutes affiliated with the Polish Academy of Sciences: Institute of Geophysics, Space Research Centre, Institute of Geological Sciences, and Institute of Oceanology, and Nicolaus Copernicus Astronomical Centre. Its main objective is a multidisciplinary approach to link scientific activities in various Earth-related fields (geophysics, geology, oceanology) with Solar System research. Our publications encompass topical monographs and selected conference proceedings, authored or edited by leading experts of international repute as well as by promising young scientists. The GeoPlanet series aims to provide the stimulus for new ideas and discoveries by reporting on the state of the art and laying the foundations for the future development of the Geosciences.

Book proposals are evaluated by the Geoplanet Editorial Office at the Institute of Geophysics PAS and Springer. Please contact Annett Buettner (annett.buettner@springer.com) for further information.

Teresa Radziejewska • Jan Marcin Węsławski
Michał Tomczak
Editors

The Natural Environment of the Mid-Atlantic Ridge

A Case Study of the Potential Mining Site

Springer

Editors
Teresa Radziejewska
Institute of Marine and Environmental
Sciences
University of Szczecin
Szczecin, Poland

Michał Tomczak
Polish Geological Institute – National
Research Institute
Warsaw, Poland

Jan Marcin Węsławski
Institute of Oceanology PAN
Sopot, Poland

ISSN 2190-5193 ISSN 2190-5207 (electronic)
GeoPlanet: Earth and Planetary Sciences
ISBN 978-3-031-51864-5 ISBN 978-3-031-51865-2 (eBook)
https://doi.org/10.1007/978-3-031-51865-2

This work was supported by Polish Geological Institute

This Springer imprint is published by the registered company Springer Nature Switzerland AG
The registered company address is: Gewerbestrasse 11, 6330 Cham, Switzerland

Acknowledgements

The review presented in this book was an initiative of the Polish Geological Institute - National Research Institute; the collection, compilation and processing of data as well as preparation of individual chapters was supported by the National Fund for the Environment and Water Management in Poland.

Executive Summary

In 2018, the Government of Poland and the International Seabed Authority (ISA) signed a contract for exploration of polymetallic sulphides in a central section of the Mid-Atlantic Ridge (MAR) (Fig. 1 in Chap. "Introduction"), one of the major geological formations on Earth (Fig. 2 in Chap. "Introduction"), featuring—like other mid-oceanic ridges—active and inactive hydrothermal vent fields with their unusual and unique ecosystems. The commercial interest in MAR exploration is due to the deposits of massive polymetallic sulphides there, formed as a result of hydrothermal fluid precipitation.

Activities associated with exploration of any area should be underpinned by knowledge on what is known about the area and its inhabitants, what gaps in knowledge still exist, what the 'rules of the game' (in the form of legal framework and regulations) are, and what methods should and need to be used to fill the gaps. Therefore, before exploration begins, the existing information on the area, including both its abiotic and biotic components, should be collected, processed, and summarized. The present book is an attempt at such a summary. The Google Scholar lists over 0.5 million scientific papers, notes, and books that mention the MAR, the relevant peer-reviewed papers counting over 60 thousand documents. For the present review, we have selected mostly recent publications dealing with the central part of the MAR, with a particular focus on the area covered by the Polish contract.

With regard to the climate conditions, the contract area is situated in the subtropics, between the easterly and westerly mid-latitude atmospheric circulation (Fig. 1 in Chap. "Hydrological and Meteorological Conditions in the Contract Area"). It is also the area on the edge of the tropical storms and hurricane tracks (Fig. 5 in Chap. "Hydrological and Meteorological Conditions in the Contract Area").

The depth of the ocean in the contract area varies between 720 m and 4900 m; the area—situated within the North Atlantic Gyre—is affected by the Gulf Stream and Canary currents. The complex seabed topography results in a number of local deep currents of smaller magnitude.

The deepwater layers are formed by the North Atlantic Deep Water with temperatures between 1.5°C and 4°C and with the full marine salinity of 34–35.

The water transparency in the area is high, with visibility reaching down to the 50 m depth, as in other mid-ocean gyres, hence the chlorophyll concentration and primary production are low.

The geological age of the seabed in the contract area is younger than 3.3 million years. The contract area features two known vent fields: the Broken Spur and the Lost City (Fig. 8 in Chap. "Seafloor Morphology, Geology, Sediments and Sedimentation Processes"). The sedimentary cover is from few to 20 m thick and is composed mainly of carbonate ooze.

The pelagic environment over the contract area supports typical oceanic microplankton assemblages, with coccolithophores and prominent contributions of diatoms and radiolarians, while the mesozooplankton is dominated by minute copepods, and is often concentrated in the sub-surface layers of 300-100 m depth. Dense concentrations of macroplankton and small fish are recorded in the mesopelagic zone. The pelagic fishery is moderately developed and targets large species such as tuna and swordfish; the commercial bottom fishing is not feasible in the area.

The pelagic vertebrate fauna of the area counts at least 60 seabird species, 28 species of whales, and four sea turtles; all those species are of conservation value, eight species being listed in the IUCN Red Book as vulnerable to critically endangered.

The organisms inhabiting the seabed in the contract area are, to large extent, adapted to the particular conditions of the hydrothermal vent environment. These organisms include chemoautotrophic and methane-consuming microbes (bacteria and archaea), the meiobenthos—an important component of the local fauna because of its diversity and potential utility as an indicator of local conditions, the macrobenthos and the megabenthos. The benthic macrofauna forms dense aggregations; particularly important are deep-sea shrimps and other macrofaunal crustaceans, locally with a record-breaking biomass exceeding 1 kg/m^2.

The endemism (uniqueness) of the deep-sea vent fauna is associated with the whole system of the MAR rather than with individual sites. This is related to a high degree of temporal stability of the benthic communities, on the scale of decades.

The conservation status of deep-sea fauna is regulated by the UN Convention on the Law of the Sea (UNCLOS), which recognizes the deep sea as the common heritage of mankind and expects any activity to observe the precautionary approach, as described in the Convention on Biological Diversity (CBD). Specific recommendations for contract-holders are issued by the ISA, and—on the next level—by the national/governmental regulations. All activities in the area shall be subject to an Environmental Impact Assessment and long-term monitoring of the biota. Methods for studying the extremely difficult deep-sea environment of the MAR involve the use of technologically advanced equipment, a strong emphasis being placed on non-invasive techniques, mainly optical and acoustic remote sensing as well as the use of robots.

Szczecin, Poland Teresa Radziejewska
Sopot, Poland Jan Marcin Węsławski

Contents

Contributors

Katarzyna Błachowiak-Samołyk Institute of Oceanology PAN, Sopot, Poland

Magdalena Błażewicz Faculty of Biology, University of Łódź, Łódź, Poland

Mateusz Damrat Polish Geological Institute – National Research Institute, Warsaw, Poland

Katarzyna Jankowska Faculty of Environmental Engineering, Technical University of Gdańsk, Gdańsk, Poland

Piotr Jóźwiak Faculty of Biology, Uniwersity of Łódź, Łódź, Poland

Bartosz Kotrys Polish Geological Institute – National Research Institute, Warsaw, Poland

Agata Kozłowska-Roman Polish Geological Institute – National Research Institute, Warsaw, Poland

Tomasz B. Linkowski Sea Fisheries Institute - National Research Institute, Gdynia, Poland

Elżbieta Maszloch Polish Geological Institute – National Research Institute, Warsaw, Poland

Krzysztof Pabis Faculty of Biology, Uniwersity of Łódź, Łódź, Poland

Jacek Piskozub Institute of Oceanology PAN, Sopot, Poland

Teresa Radziejewska Institute of Marine and Environmental Sciences, University of Szczecin, Szczecin, Poland

Sławomir Sagan Institute of Oceanology PAN, Sopot, Poland

Łukasz Smajdor Polish Geological Institute – National Research Institute, Warsaw, Poland

Lech Stempniewicz Faculty of Biology, University of Gdansk, Gdańsk, Poland

Paweł Sydor Polish Geological Institute – National Research Institute, Warsaw, Poland

Witold Szczuciński Faculty of Geography & Geology, Adam Mickiewicz University, Poznań, Poland

Agata Szyduk Polish Geological Institute – National Research Institute, Warsaw, Poland

Michał Tomczak Polish Geological Institute – National Research Institute, Warsaw, Poland

Waldemar Walczowski Institute of Oceanology PAN, Sopot, Poland

Jan Marcin Węsławski Institute of Oceanology PAN, Sopot, Poland

Maria Włodarska-Kowalczuk Institute of Oceanology PAN, Sopot, Poland

Tomasz Żuk Polish Geological Institute – National Research Institute, Warsaw, Poland

Geoscience FreeD - Open Geomodelling, Warsaw, Poland

Introduction

Teresa Radziejewska, Jan Marcin Węsławski, and Mateusz Damrat

The Polish interest in the study area is reflected by the contract for polymetallic sulphide exploration between the International Seabed Authority (ISA) and the Ministry of Environment of the Republic of Poland, signed in 2018. The Polish exploration contract area is located in the central part of the Atlantic Ocean (Fig. 1) between 26°09′N–32°50′N and 39°38′W–44°43′W near the central part of the Mid-Atlantic Ridge (MAR), one of the major geological formations on Earth (Fig. 2), featuring—like other mid-oceanic ridges—active and inactive hydrothermal vent fields with their unusual and unique ecosystems. The water depth varies from about 740 m above the MAR to more than 4900 m in fracture zones and basins. The Polish exploration blocks are located to the south-west of the Portuguese submission to the Commission on the Limits of the Continental Shelf and to the north-east of the French exploration contract area. The contract area covers 10,000 km^2 and consists of 100 exploration blocks, 10 × 10 km each. The blocks are grouped in 5 clusters (Fig. 1).

In recent decades, the MAR has received extreme attention both for its scientific value and potential commercial interest. The commercial interest in MAR exploration is due to the deposits of polymetallic massive sulphides (PMS), formed as a result of hydrothermal fluid precipitation. Those deposits are considered a strategic

T. Radziejewska
Institute of Marine and Environmental Sciences, University of Szczecin, Szczecin, Poland
e-mail: teresa.radziejewska@usz.edu.pl

J. M. Węsławski (✉)
Institute of Oceanology PAN, Sopot, Poland
e-mail: weslaw@iopan.gda.pl

M. Damrat
Polish Geological Institute – National Research Institute, Warsaw, Poland
e-mail: mdam@pgi.gov.pl

T. Radziejewska et al. (eds.), *The Natural Environment of the Mid-Atlantic Ridge*, GeoPlanet: Earth and Planetary Sciences,
https://doi.org/10.1007/978-3-031-51865-2_1

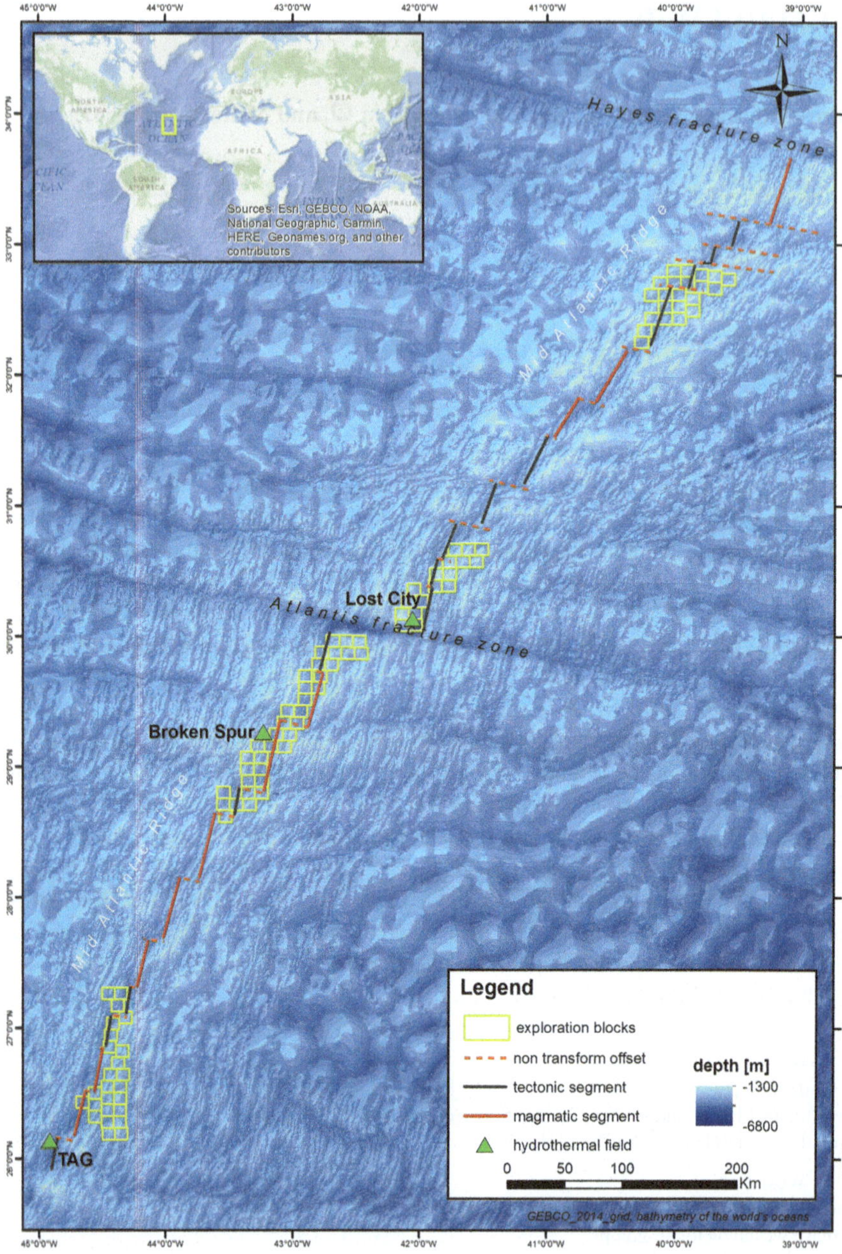

Fig. 1 The Polish contract area on the Mid-Atlantic Ridge (MAR) divided into 5 clusters (Kozłowska-Roman et al. 2019)

Fig. 2 The Mid-Atlantic Ridge with its seabed relief: a fragment of the National Geographic seabed map of 1998

mineral resource of the future. As such, and because most of the MAR lies beyond borders of national jurisdictions, the MAR seafloor and its resources—as part of the common heritage of mankind—are managed by the International Seabed Authority (ISA), an international entity entitled to conclude contracts for exploration of those resources with suitably qualified parties (contractors). In 2018, Poland concluded such a contract and has recently commenced exploration.

The main objective of the contract is the identification of hydrothermally-based PMS and their economic evaluation. Hydrothermal sulphides contain mainly Cu, Zn, Fe and smaller concentrations of Au, Ag, Co, Ni, Cd, In, REE and other elements. As the seabed spreads, the active hydrothermal areas are gradually cut off from the energy source and become inactive (extinct or dormant); these, in contrast to hydrothermally active zones, are known much less. The identification of hydrothermally inactive sites is more difficult and is based on the imaging of the seabed with high-resolution multibeam echo sounders, cameras and the detection of negative self (electrical) potential anomalies. Despite these limitations, the total mass of polymetallic sulphide deposits in hydrothermally inactive areas is estimated at 600 Mt., 20 times more than in the recognized deposits on land; thus, the ridge deposits form areas of potential massive sulphide exploitation.

An integral part of exploration involves comprehensive environmental baseline studies which should aid the assessment of a potential impact of the future exploitation.

Currently, most evidence on depositional conditions, environmental parameters and biodiversity of the ridge areas is provided by research conducted under research (exploration) contracts, issued under agreements with the ISA, established in 1982 under the United Nations Convention on the Law of the Sea (UNCLOS). The contracts provide a stable platform for comprehensive oceanographic, biological and geological research carried out according to strictly defined rules, under the control of the ISA and in cooperation at the national and international level.

Thanks to the possibility of exploring the unique environment of the MAR, Polish scientists will have a chance to contribute new data and research results to the body of knowledge on the MAR, generated by the international scientific community.

Reference

Kozłowska-Roman A, Tomczak M, Skowronek A (2019) Characteristic of the Polish reserved exploration area in the Mid-Atlantic Ridge. In: 48th Underwater Mining Conference, Sanya, China, September 22–27, 2019. Abstract book, pp 112–118

Part I
Abiotic Environment

Hydrological and Meteorological Conditions in the Contract Area

Jacek Piskozub, Waldemar Walczowski, and Sławomir Sagan

1 Meteorology and Air-Sea Interaction

1.1 General Atmospheric Circulation

The Polish contract area on the Mid Atlantic Ridge (MAR) between 26°09′–32°50′N; 39°38′–44°43′W is technically placed in the subtropics defined usually as having the northern boundary at the 35°N latitude (AMS 2020a, b). However, in practice the area is situated on the border between the easterly subtropical circulation (trade winds) of the subtropics and the westerly circulation of the middle latitudes (Fig. 1). The actual position of the border changes seasonally, moving northward during the (boreal) warm seasons and southward during cold ones (Holton and Hakim 2012).

Figure 2 shows the average zonal (east-west) wind component for 10 years (2010–2019) for each season, based on data from the NCEP/NCAR reanalysis (Kalnay et al. 1996). The data show the northern part of the contract area to be within the Westerlies zone during the winter and spring seasons (more precisely from December to June, not shown). The subtropical area of trade winds is generally a region of steady and predictable winds, but the Westerlies area is dominated by extratropical cyclonic (storms) and anticyclonic systems driven by atmospheric Rossby waves (Rhines 2015). Figure 3 shows the average wind speeds for each season in the region. The seasonal (and monthly, not shown) averages are between 6 and 8 m/s, except for the northernmost part where the average wind speed reaches 9 m/s in winter. North of the contract area, the average wind speed increases due to the region being situated in the Atlantic storm track. In the warming world, a slow northward shift of the Atlantic storm tracks is both observed (Bender et al. 2012)

J. Piskozub (✉) · W. Walczowski · S. Sagan
Institute of Oceanology PAN, Sopot, Poland
e-mail: piskozub@iopan.gda.pl; walczows@iopan.gda.pl; sagan@iopan.gda.pl

T. Radziejewska et al. (eds.), *The Natural Environment of the Mid-Atlantic Ridge*, GeoPlanet: Earth and Planetary Sciences,
https://doi.org/10.1007/978-3-031-51865-2_2

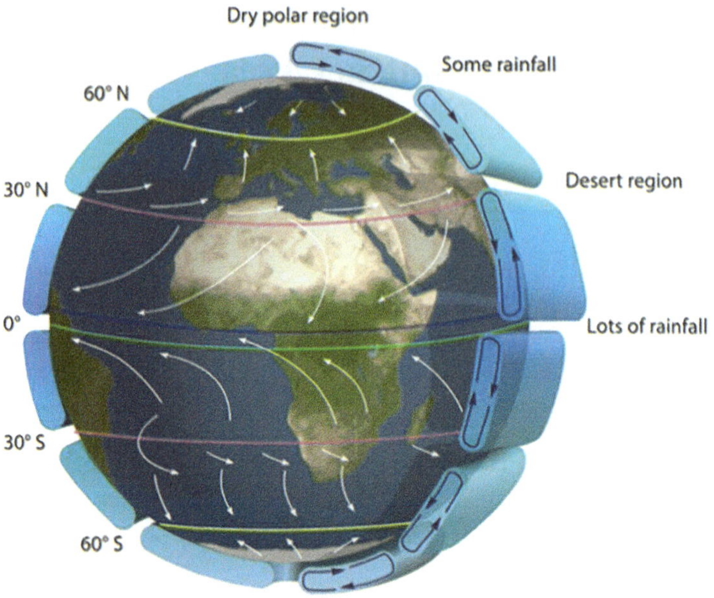

Fig. 1 Global atmospheric circulation showing the border between easterly (trade winds) and westerly circulation close to 30°N (source: Wikipedia)

and projected (Yin 2005), but the trends are too inconspicuous to expect that the storm track will move away from the contract area in the near future.

1.2 Extreme Weather Events

The mean values of hydrographic parameters obscure extreme events such as winter storms in the northern part of the contract area as well as summer hurricanes. The maximal sustained 6-h values of wind are shown in Fig. 4. The values of the maximal sustained wind in the license area are lowest (14–18 m/s) in summer (June–August) and highest (20–25 m/s) in winter (December–February). As the values are averages for 2.5° × 2.5° areas, local values are expected to be even higher. In particular, tropical cyclones (called hurricanes in the North Atlantic basin) are too small for their maximum wind speed to be fully registered with the spatial averaging used in the NCEP/NCAR reanalysis. However, it is evident from Fig. 4 that several hurricane tracks appear on the autumn (September–November) map, the closest track occurring just south of the contract area. That hurricane was the 2019 Lorenzo which reached its peak intensity of 140 kt (72 m/s) on September 29, 2019, just south of the contract area (Zelinsky 2019), to become briefly a Category 5 hurricane on the Saffir-Simpson scale (AMS 2020a, b). It moved northward, still with a hurricane strength, along the westerly edge of the contract area. Even if there was only

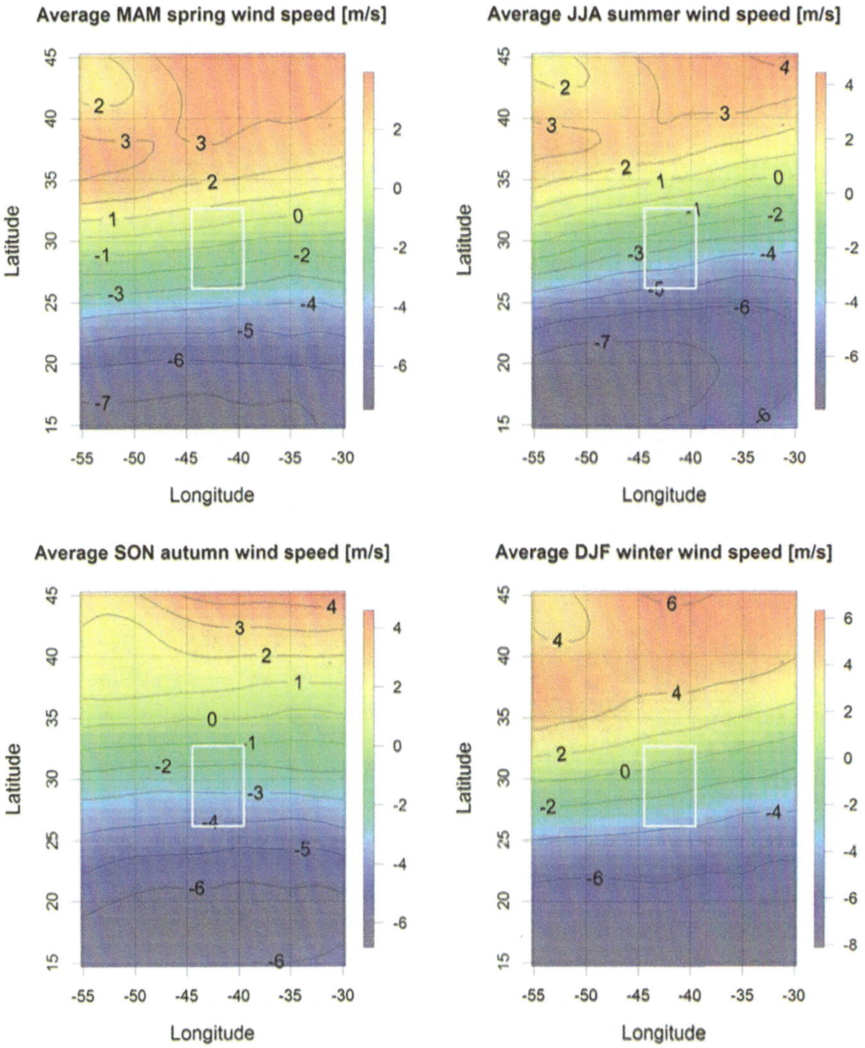

Fig. 2 Average zonal (east-west) component of the surface winds in the central North Atlantic by season. Data for 2010–2019 based on the NCEP/NCAR reanalysis. Positive numbers denote westerly winds. The Polish contract area is marked with a white rectangle (MAM, March–April-May; JJA, June–July-August; SON, September–October-November; DJF, December–January-February)

one hurricane moving over the contract area during the last 30 years (1990–2019), this shows that such a possibility exists and has to be taken into account.

A review of the whole HURDAT (Landsea and Franklin 2013) tropical cyclone climatology (1851–2015) shows a total of 26 tracks of named tropical storms and hurricanes crossing the contract area, i.e., 1.5 per decade, although this number needs to be treated as only approximate as the storm tracks in the pre-aircraft era

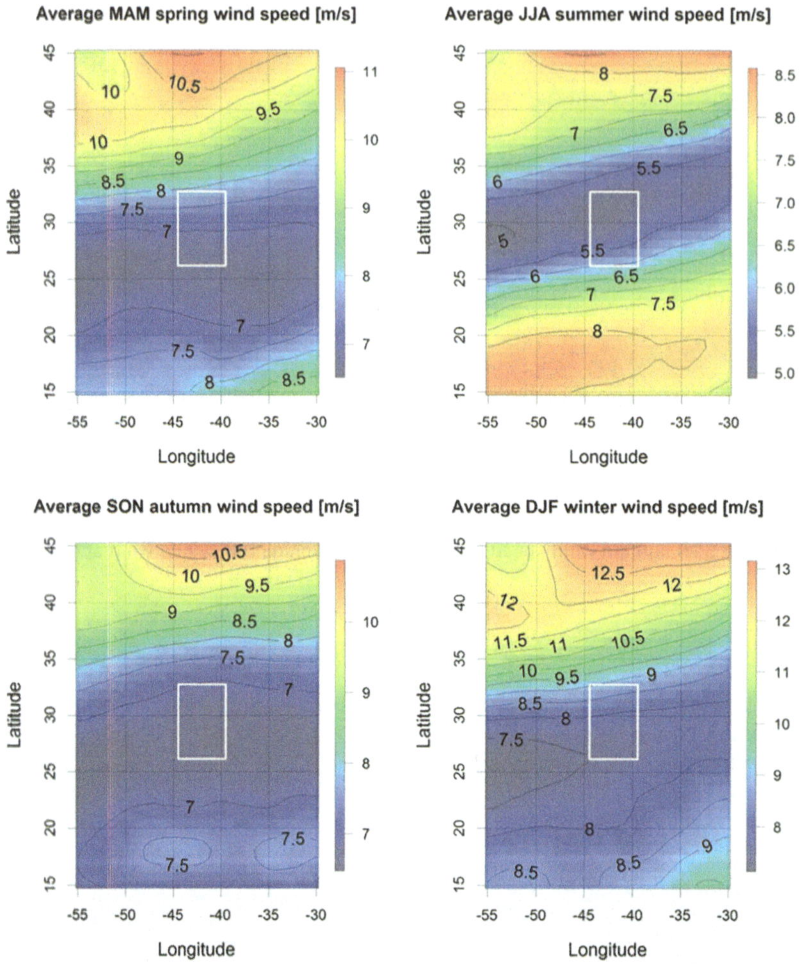

Fig. 3 Average surface (U10) wind speed in the central North Atlantic by season for 2010–2019, based on the NCEP/NCAR reanalysis. The Polish contract area is marked with a white rectangle. See Fig. 2 for explanation of season symbols

(before 1945) are known only from re-analytical reconstructions based on few observational data. Figure 5 shows all the reconstructed tropical storm and hurricane tracks for 1851–2015. It is evident that hurricanes in the region of the contract area usually move from south to north. This is caused by the so-called beta effect, i.e., the force exerted on a tropical cyclone by the meridional gradient of the Coriolis force (Chan and Williams 1987), whereby the system is being pushed northward, towards the pole. The U-shaped tracks visible in Fig. 5 are the result of the beta effect working together with the easterly circulation south of 30°N and the westerlies north of it. As the contract area is at (or very close to) the latitude of zero zonal

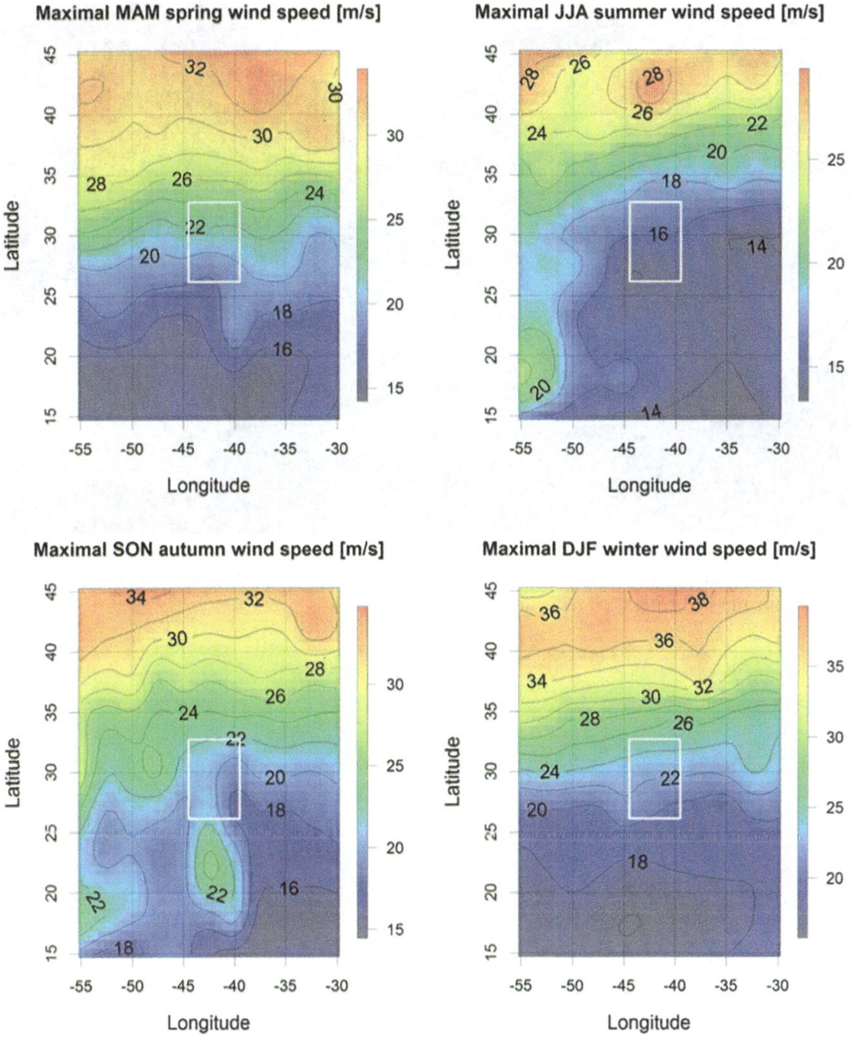

Fig. 4 Maximal 6-h sustained values of surface (U10) wind speed in the central North Atlantic by season for 1990–2019, based on the NCEP/NCAR reanalysis. The Polish contract is marked with a white rectangle. See Fig. 2 for explanation of season symbols

circulation in late summer and early autumn, the only significant force working on the cyclone is the beta effect. As a result, the hurricanes threatening the contract area need to be generated south of it or within the area itself. It is well known that tropical cyclones develop and are sustained only over those ocean waters of a sea surface temperature (SST) higher than 27 °C (various authors refer to a range 25.5 to 27 °C). However, the strongest (Category 4 and 5) hurricanes are sustained only above

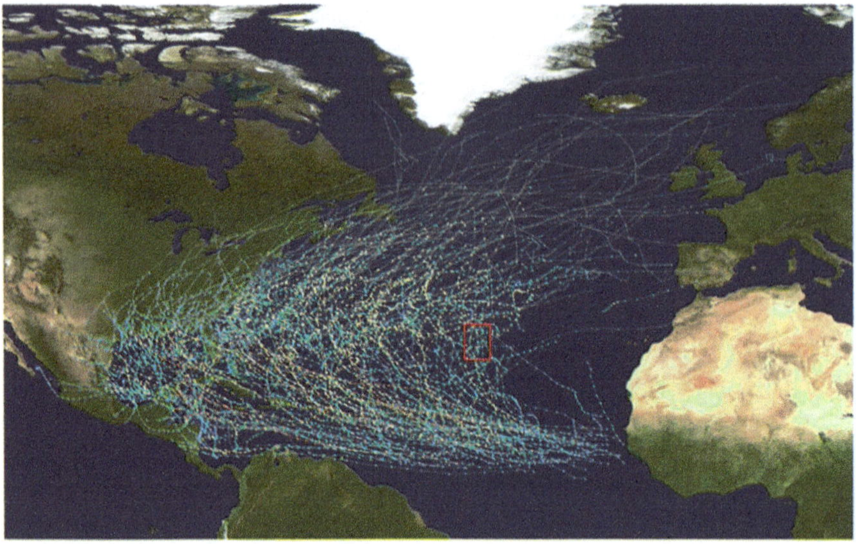

Fig. 5 Hurricane and tropical storm tracks, 1851–2015. Approximate location of the Polish contract area is marked with a red rectangle (source: Wikipedia)

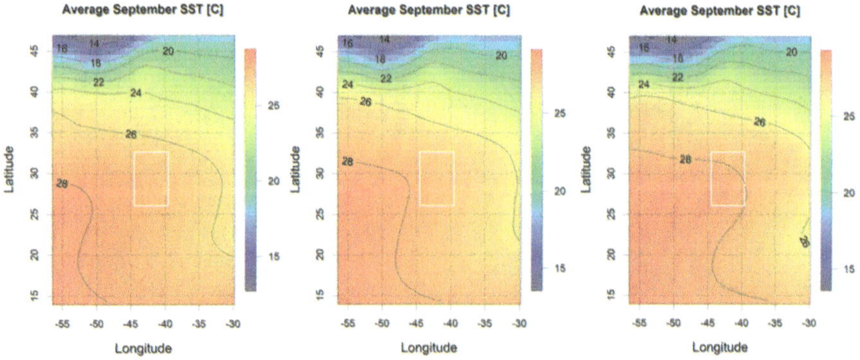

Fig. 6 Average September sea surface temperature (SST) in the central North Atlantic from the NCEP/NCAR reanalysis; left: 1981–2010; centre: 2010–2019; right: 2019; the Polish contract area is marked with a white rectangle

water with SST of 28 °C or more (Elsner et al. 2013). Figure 6 shows the September average SST in the central part of North Atlantic for three periods: 1981–2010, 2010–2019, and the year 2029. It can be seen that the 28 °C isotherm has, with time, shifted towards the contract area, rendering it warmer than the threshold in 2019, which may explain why the hurricane Lorenzo visited the area at the end of September. With global warming, climate models (Haarsma et al. 2013) predict such extreme meteorological events to be become increasingly frequent in the area.

1.3 Atmospheric Pollution

The Polish contract area lies in the central part of the North Atlantic in a region far away from any continent. The nearest land are the Azores. The upwind position of the archipelago makes it unlikely that any pollution from the archipelago reaches the contract area in question, except for periods in autumn and winter when the westerlies (storms) pass over its northern part. With a typical trade-wind circulation, the closest land areas downwind are the Caribbean islands and South America. The same circulation may bring the Saharan dust plume to the contract area, especially in summer (Gläser et. al, 2015).

2 Hydrography

The Polish contract area is located in the subtropical zone of the North Atlantic. The Mid-Atlantic Ridge (MAR) extends across the centre of the area, from the southwest to the northeast. In the region, the highest underwater elevations rise to a depth less than 800 m, the deepest parts being 4900 m deep (EMODnet 2018). The MAR is intersected by the Atlantis and Hayes fracture zones

2.1 Oceanic Circulation

Due to the shape of the Atlantic Ocean, its meridional extension and open, broad connections with the Southern and the Arctic oceans, the circulation is dominated by the large-scale Atlantic Meridional Overturning Circulation (AMOC) (Kuhlbrodt et al. 2007). In the surface layer, the AMOC carries warm, salty water towards the northern and southern polar regions where it cools down, sinks to deep water column layers, and returns towards the equator as dense, cold deep waters. The surface currents (Fig. 7), forced mainly by the wind and thermohaline effects, are more intense and concentrated than the intermediate and deep ones. The temporal, short-term variability of the AMOC is maintained by the Ekman transport; the thermohaline geostrophic component which reflects the east-west density contrasts across the ocean, dominates on interannual to decadal timescales (Buckley and Marshall 2016)

In the northern hemisphere, the Coriolis force results in a clockwise circulation and concentration of the strongest currents at the western side of the subtropical Atlantic Ocean. The surface northward water transport is dominated by the Gulf Stream. North of the 40° N parallel, the Gulf Stream splits into the northward-flowing North Atlantic Current and the eastward-propagating Azores Current (Tomczak and Godfrey 2001). The Portugal and Canary currents carry the water southward along the southern Europe and the western coasts of Africa. The anticyclonic loop of the North Atlantic Gyre (NAG) is closed by the North Equatorial

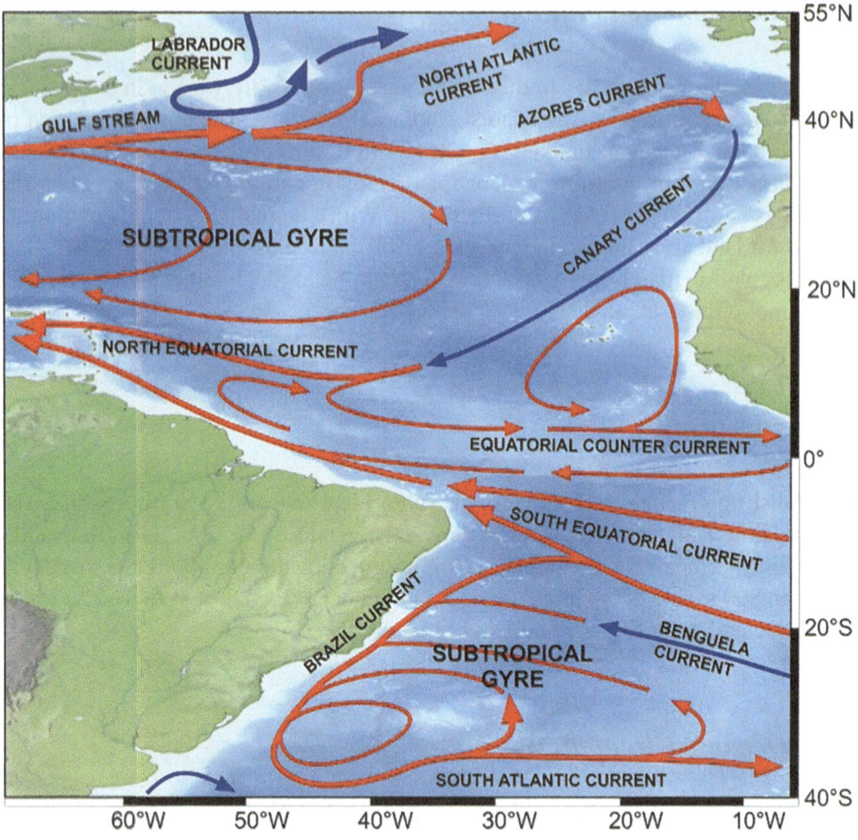

Fig. 7 The current pattern in the region

Current flowing westward between latitudes 10°N and 20°N. The maximum current velocities in the surface layer exceed 100 cm/s, the climatic averaged transports being an order of magnitude slower. As the region of the Polish contract area is situated in the centre of the NAG, the surface currents sustained by inner, southward flowing branches are less intensive there. The intermediate water pattern is similar to the surface flows, but the velocities are much lower, on the order of a few cm/s.

The deep circulation of the Atlantic Ocean is dominated by the Deep Western Boundary Current carrying cold, dense water southward from the Arctic. The presence of the MAR results in numerous local deep currents, steered by the seabed topography. They carry the water both along the MAR and across it, through underwater canyons and rift valleys. All these flows contribute significantly to the enhanced vertical mixing of the deep water (Mauritzen et al. 2002). The turbulent, diapycnal mixing in rift valleys plays an especially important role (Thurnherr et al. 2002). Tidal processes enhanced by bottom topography increase mixing as well. Horizontal and vertical flows associated with hydrothermal vent plumes are also observed (Thurnherr et al. 2008).

2.2 *Water Masses*

The distinct layers visible in the circulation pattern result in a general division of water masses (Figs. 8 and 9) into the surface (0–500 m), intermediate (500–1500 m), and deep (1500-bottom) waters (Emery 2001). Each region features more complex sub-divisions. In general, the surface waters are subject to seasonal variability in solar radiation and atmospheric forcing. Therefore, properties of these waters (temperature, salinity, density) undergo the highest seasonal variability. In addition, a distinct spatial variability is present as well. Argo data from the region over the MAR in the Polish contract area have revealed temperature and salinity ranges of 20.5–28.0 °C and 36.4–37.8, respectively. The intermediate waters are less variable: the temperature and salinity ranges for the Western Atlantic Subarctic Intermediate Water (WASIW) are 3.0–9.0 °C and 34.0–35.1, respectively, the respective ranges for the Eastern Atlantic Subarctic Intermediate Water (EASIW) being 3.0–9.0 °C and 34.4–35.3.

Dominant is the regional variability, with no distinct seasonal variations. The most stable are the properties of the deep waters. The dominant abyssal water mass in the MAR region is the North Atlantic Deep Water (NADW) with mean temperature and salinity of 1.5–4.0 °C and 34.0–35.0, respectively. As NADW is formed at

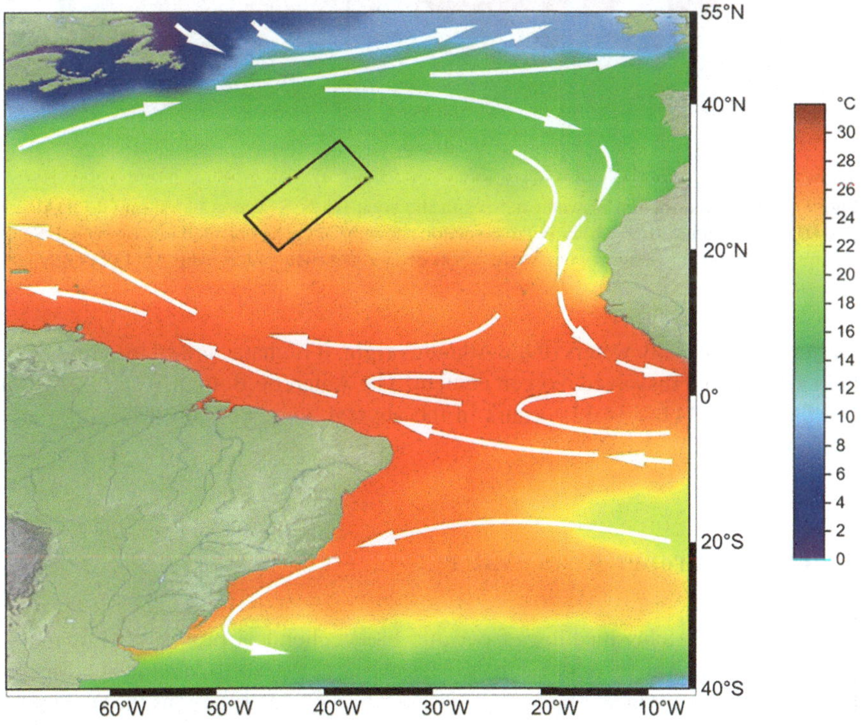

Fig. 8 Sea surface temperature in summer

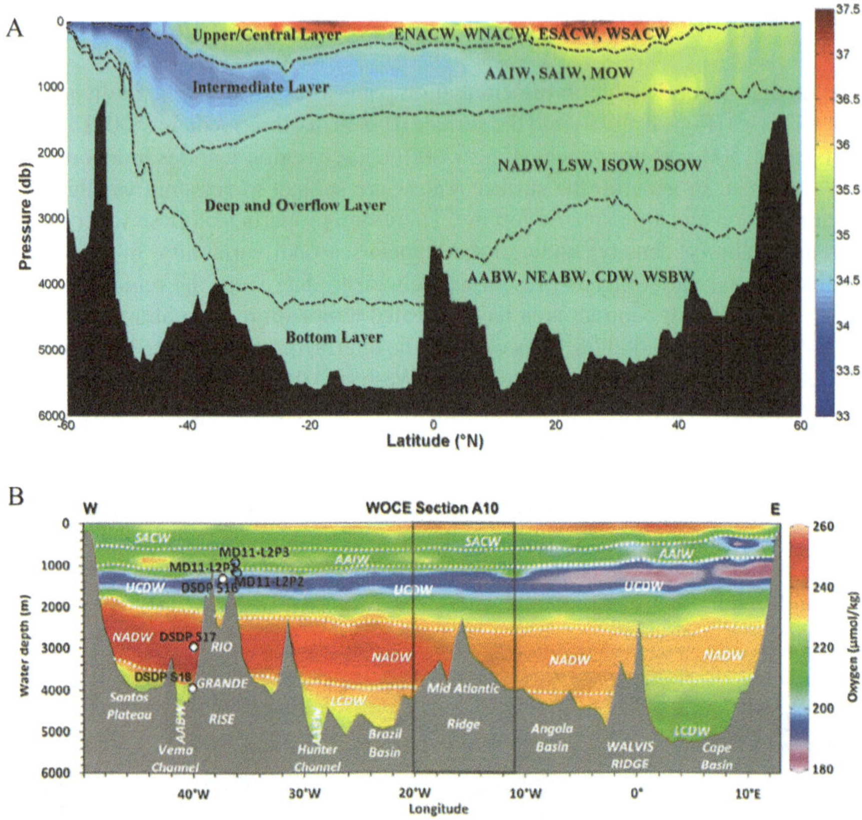

Fig. 9 Vertical distribution of water masses in the area: (**a**) N-S section based on GLODAP v2 data (Liu and Tanhua 2019); (**b**) E-W cross section showing oxygen content, based on WOCE Sect. 5. https://www.nodc.noaa.gov/woce/woce_v3/wocedata_2/ewoce/gallery/whp/A5_OXYGEN.gif

the surface from salty waters of the Atlantic origin, it is characterised by a relatively high salinity and dissolved oxygen content. The Antarctic Bottom Water (AABW) (−0.9–1.7 °C; 34.64–34.72) occurs in the abyssal zone, both on the eastern and western sides of the MAR.

3 Optical Properties of Water

According to the classification of oceanic biogeochemical provinces (Longhurst 2007), the contract area is located in the central part of the North Atlantic Subtropical Gyre (NASTG) province. The optical classification of waters assigns the area to the so-called 'Case 1' waters, i.e., the waters whose optical properties depend on the chlorophyll content, with a small or insignificant influence of non-organic particles and chromophoric dissolved organic matter (CDOM) (Morel and Prieur 1977). The concurrent processes of light scattering and absorption by water constituents and molecules of water result here in the distinctive blue colour of water and its high transparency, with Secchi disc visibility reaching 50 m (Longhurst 2007). Consequently, the euphotic zone, defined as a depth of 1% of surface solar radiation, reaches down to the 150–200 m depth. Due to the high water transparency, light is detectable down to the bottom of the mesopelagic zone (1000 m), and its colour is centred around the blue light of 485 nm wavelength (Li et al. 2014)

However, compared to the clearest world oceans, the NASTG is considerably less blue than other oligotrophic subtropical zones. The area is characterized by a CDOM content higher than the ocean average, with the light absorption coefficient aCDOM (380 nm) of 0.015–0.025/m (Organelli and Claustre 2019). This enhanced light absorption in the blue region of the light spectrum is attributed to an elevated local CDOM production by the picophytoplankton, believed to increase in importance under the ocean warming scenario (Organelli and Claustre 2019). The other water constituent affecting optical properties, i.e., suspended particles, remains low. The light scattering coefficient, bp (526 nm), a proxy for particle concentration, varies from 0.03 to 0.05/m, which is typical of oligotrophic oceanic waters (Dall'Olmo et al. 2012)

Primary production, chlorophyll contents, and optical properties of the surface water in the area are assessed based on a combination of satellite observations with *in situ* data, acquired mainly during several expeditions of the Atlantic Meridional Transect (AMT) programme (Longhurst 2007; Tilstone et al. 2009; Brewin et al. 2017; Rees et al. 2017; Tilstone et al. 2017). Primary production varies with the seasonal cycles of chlorophyll contents in and changes of the mixed-layer depth (Fig. 10). The monthly averaged depth-integrated primary production varies from 0.15 to 0.65 gC/m^2 d, the chlorophyll concentration ranging within 0.05–0.25 mg/m^3. The euphotic zone thickness remains relatively stable and varies seasonally between 48 and 52 m. Primary producers in the area are dominated by the smallest phytoplankton size spectra, and the multiyear trend points to a relatively stable primary production (Gregg et al. 2003; Varela et al. 2005; Vicente et al. 2012)

Optical properties of the deep ocean waters depend mainly on the presence of sinking particles, the uniform spatial distribution of which constitutes the background level of the light scattering coefficient similar to that in the upper layer, i.e., around 0.05/m or—in terms of turbidity units—~40 NTU (Murton et al. 1994). However, optical properties of the water 50–500 m above the bottom might be affected by the buoyant plumes from hydrothermal vents, which transport particulate material (Fig. 11). Hydrothermal plumes can spread laterally over distances of a few to several hundreds of kilometres (Resing et al. 2015). Contrasting with the transparent, virtually particle-free background water, the plumes showed turbidity values of 200 NTU, as the light attenuation coefficient in the vicinity of vents exceeded 100/m (Estapa et al. 2015); typical values recorded in coastal waters considered turbid are 5–20/m.

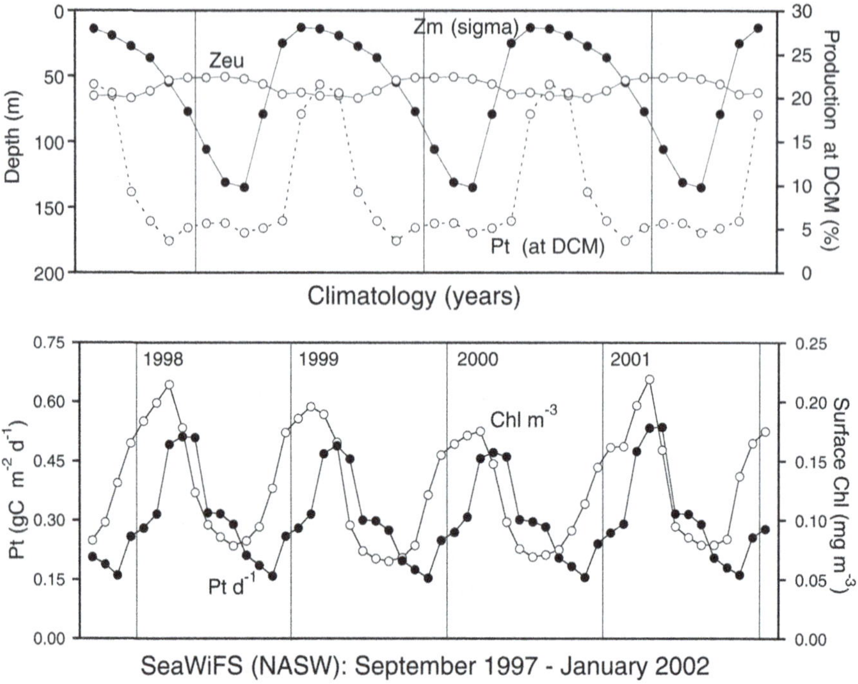

Fig. 10 Seasonal cycles of monthly surface chlorophyll (Chl) and depth-integrated autotrophic production (Pt) in the North Atlantic Subtropical Gyre (NASTG) province for 1997–2002 from SeaWiFS data together with characteristic seasonal cycles of mixed-layer depths (Zm), euphotic zone (Zeu) and Pt at the Deep Chlorophyll Maximum depth (Longhurst 2007)

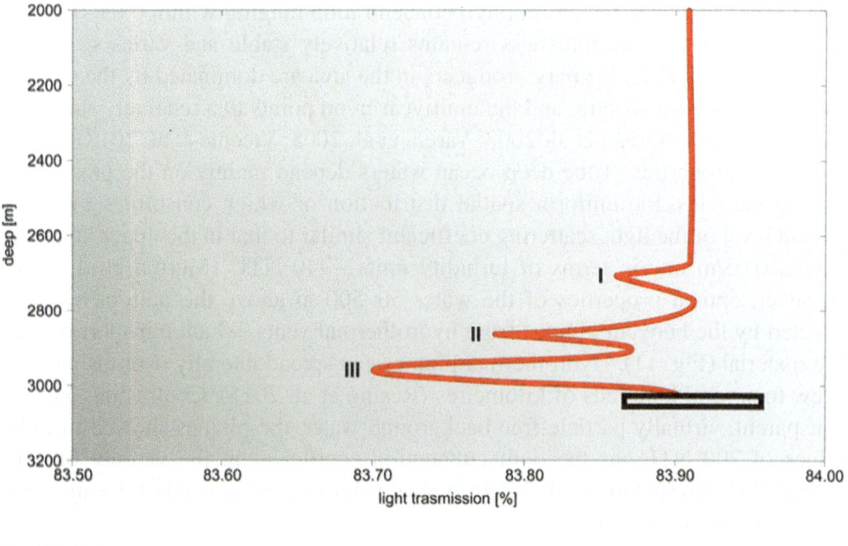

Fig. 11 Evidence of the lateral transport of a particle plume from a hydrothermal vent seen as a drop of the light transmission (Murton et al. 1994)

References

American Meteorological Society (2020a) "Saffir–Simpson hurricane scale". Glossary of Meteorology. https://glossary.ametsoc.org/wiki/Saffir-impson_hurricane_scale

American Meteorological Society (2020b) "Subtropics". Glossary of Meteorology. https://glossary.ametsoc.org/wiki/Subtropics

Bender FAM, Ramanathan V, Tselioudis G (2012) Changes in extratropical storm track cloudiness 1983–2008: observational support for a poleward shift. Clim Dyn 38:2037–2053

Brewin RJW, Tilstone GH, Jackson T, Cain T, Miller PI, Lange PK, Misra A, Airs RL (2017) Modelling size-fractionated primary production in the Atlantic Ocean from remote sensing. Progress in Oceanography, The Atlantic Meridional Transect programme 1995-2016(158):130-149. https://doi.org/10.1016/j.pocean.2017.02.002

Buckley MW, Marshall J (2016) Observations, inferences, and mechanisms of the Atlantic meridional overturning circulation: a review. Rev Geophys 54:5–63

Chan JCL, Williams RT (1987) Analytical and numerical studies of the beta-effect in tropical cyclone motion. Part I: zero mean flow. J Atmos Sci 44:1257–1265

Dall'Olmo G, Boss E, Behrenfeld MJ, Westberry TK (2012) Particulate optical scattering coefficients along an Atlantic meridional transect. Opt Express 20:21532–21551. https://doi.org/10.1364/OE.20.021532

Elsner JB, Strazzo SE, Jagger TH, LaRow T, Zhao M (2013) Sensitivity of limiting hurricane intensity to SST in the Atlantic from observations and GCMs. J Clim 26:5949–5957

Emery WJ (2001) Water types and water masses. In: Steele JH (ed) Encyclopedia of Ocean Sciences. Academic, New York, pp 3179–3187

EMODnet Bathymetry Consortium (2018) EMODnet Digital Bathymetry (DTM). https://doi.org/10.12770/18ff0d48-b203-4a65-94a9-5fd8b0ec35f6

Estapa ML, Breier JA, German CR (2015) Particle dynamics in the rising plume at Piccard hydrothermal field, Mid-Cayman Rise. Geochem Geophys Geosyst 16:2762–2774. https://doi.org/10.1002/2015GC005831

Gläser G, Wernli H, Kerkweg A, Teubler F (2015) The transatlantic dust transport from North Africa to the Americas—its characteristics and source regions. J Geophys Res Atmos 120:11231–11252

Gregg WW, Conkright ME, Ginoux P, O'Reilly JE, Casey NW (2003) Ocean primary production and climate: global decadal changes. Geophys Res Lett 30:1809. https://doi.org/10.1029/2003GL016889

Haarsma RJ, Hazeleger W, Severijns C, de Vries H, Sterl A, Bintanja A, van Oldenborgh GJ, van den Brink HW (2013) More hurricanes to hit western Europe due to global warming. Geophys Res Lett 40:1783–1788

Holton J, Hakim GJ (2012) The general circulation. In: Holton J, Hakim GJ (eds) An introduction to dynamic meteorology, 5th edn. Academic, San Diego, CA, pp 325–375

Kalnay E, Kanamitsu M, Kistler R, Collins W, Deaven D, Gandin L, Iredell M, Saha S, White G, Woollen J, Zhu Y, Chelliah M, Ebisuzaki W, Higgins W, Janowiak J, Mo KKC, Ropelewski C, Wang J, Leetmaa A, Reynolds R, Jenne R, Joseph D (1996) The NCEP/NCAR 40-year reanalysis project. Bull Am Meteorol Soc 77(3):437–471. https://doi.org/10.1175/1520-0477(1996)077<0437:TNYRP>2.0.CO

Kuhlbrodt T, Griese A, Montoya M, Levermann A, Hofmann M, Rahmstorf S (2007) On the driving processes of the Atlantic meridional overturning circulation. Rev Geophys 45:2004RG000166. https://doi.org/10.1029/2004RG000166

Landsea CW, Franklin JL (2013) Atlantic hurricane database uncertainty and presentation of a new database format. Mon Weather Rev 141:3576–3592

Li L, Stramski D, Reynolds RA (2014) Characterization of the solar light field within the ocean mesopelagic zone based on radiative transfer simulations. Deep Sea Res Part I Oceanogr Res Pap 87:53–69. https://doi.org/10.1016/j.dsr.2014.02.005

Liu M, Tanhua T (2019) Distribution of water masses in the Atlantic Ocean based on 1 GLODAPv2. Ocean Science Discussions. https://doi.org/10.5194/os-2018-140

Longhurst AR (2007) Ecological geography of the sea.2nd edn. Academic, Amsterdam

Mauritzen C, Polzin KL, McCartney MS, Millard RC, West-Mack DE (2002) Evidence in hydrography and density fine structure for enhanced vertical mixing over the Mid-Atlantic Ridge in the western Atlantic. J Geophys Res 107(C10):3147. https://doi.org/10.1029/2001JC001114

Morel A, Prieur L (1977) Analysis of variations in ocean color. Limnol Oceanogr 22:709–722. https://doi.org/10.4319/lo.1977.22.4.0709

Murton BJ, Klinkhammer G, Becker K, Briais A, Edge D, Hayward N, Millard N, Mitchell I, Rouse I, Rudnicki M, Sayanagi K, Sloan H, Parson L (1994) Direct evidence for the distribution and occurrence of hydrothermal activity between 27°N–30°N on the Mid-Atlantic Ridge. Earth Planet Sci Lett 125:119–128. https://doi.org/10.1016/0012-821X(94)90210-0

Organelli E, Claustre H (2019) Small phytoplankton shapes colored dissolved organic matter dynamics in the North Atlantic subtropical gyre. Geophys Res Lett 46:12183–12191. https://doi.org/10.1029/2019GL084699

Rees AP, Nightingale PD, Poulton AJ, Smyth TJ, Tarran GA, Tilstone GH (2017) The Atlantic Meridional Transect Programme (1995–2016). Prog Oceanogr 158:3–18. https://doi.org/10.1016/j.pocean.2017.05.004

Resing JA, Sedwick PN, German CR, Jenkins WJ, Moffett JW, Sohst BM, Tagliabue A (2015) Basin-scale transport of hydrothermal dissolved metals across the South Pacific Ocean. Nature 523:200–203. https://doi.org/10.1038/nature14577

Rhines PB (2015) Rossby waves. In: North GR, Pyle J, Zhang F (eds) Encyclopaedia of Atmospheric Sciences, vol 2. Academic, San Diego, CA, pp 404–416

Thurnherr AM, Richards KJ, German CR, Lane-Serff GF, Speer KG (2002) Flow and mixing in the rift valley of the Mid-Atlantic Ridge. J Phys Oceanogr 32:1763–1778

Thurnherr AM, Reverdin G, Bouruet-Aubertot P, St. Laurent LC, Vangriesheim A, Ballu V (2008) Hydrography and flow in the Lucky Strike segment of the Mid-Atlantic Ridge. J Mar Res 66:347–372

Tilstone G, Smyth T, Poulton A, Hutson R (2009) Measured and remotely sensed estimates of primary production in the Atlantic Ocean from 1998 to 2005. Deep Sea Res Part II Top Stud Oceanogr 56:918–930. https://doi.org/10.1016/j.dsr2.2008.10.034

Tilstone GH, Lange PK, Misra A, Brewin RJW, Cain T (2017) Micro-phytoplankton photosynthesis, primary production and potential export production in the Atlantic Ocean. Prog Oceanogr 158:109–129. https://doi.org/10.1016/j.pocean.2017.01.006

Tomczak M, Godfrey JS (2001) Regional oceanography: an introduction. Pergamon Press

Varela MM, Bode A, Fernandez E, Gonzalez N, Kitidis V, Varela M, Woodward EMS (2005) Nitrogen uptake and dissolved organic nitrogen release in planktonic communities characterised by phytoplankton size-structure in the Central Atlantic Ocean. Deep Sea Res Part I Oceanogr Res Pap 1(52):1637–1661

Vicente VM, Tilstone GH, Sathyendranath S, Miller PI, Groom SB (2012) Contributions of phytoplankton and bacteria to the optical backscattering coefficient over the Mid-Atlantic Ridge. Mar Ecol Prog Ser 445:37–51

Yin JH (2005) A consistent poleward shift of the storm tracks in simulations of 21st century climate. Geophys Res Lett 32:L18701

Zelinsky DA (2019) Hurricane Lorenzo (AL 132019), National Hurricane Center Tropical Cyclone Report. https://www.nhc.noaa.gov/data/tcr/AL132019_Lorenzo.pdf

Seafloor Morphology, Geology, Sediments and Sedimentation Processes

Paweł Sydor, Witold Szczuciński, Tomasz Żuk, and Agata Kozłowska-Roman

1 Introduction

The mid-ocean ridges form a submarine mountain range (Fig. 1) and are considered to be the largest geological feature on Earth. Their total length is approximately 5000 km, while they are about 1000 km wide. Altogether they cover ca. 23% of Earth's surface (Earle 2020).

The mid-ocean ridges are located at divergent oceanic plate boundaries, where magmatic and volcanic processes associated with seafloor spreading generate new oceanic crust. The rate of seafloor spreading determines the morphology of the crest of a mid-ocean ridge and its width, with higher spreading rates producing a more gentle and wider ridge (e.g., in the Pacific Ocean) and lower spreading rates resulting in a more pronounced, narrower ridge (e.g., in the Atlantic Ocean; Fig. 1). The morphology of mid-ocean ridges is a result of interplay of volcanic, tectonic and sedimentary processes. In the central part, there is usually a well-visible rift valley (Fig. 2). The ridge is dissected by multiple faults: small-scale faults forming local-scale complex topography (e.g., Fig. 3) and large transform faults/fracture zones (Fig. 1), perpendicular to the ridge axis. The accumulation of sediments (mainly

P. Sydor · A. Kozłowska-Roman
Polish Geological Institute – National Research Institute, Warsaw, Poland

W. Szczuciński (✉)
Institute of Geology, Adam Mickiewicz University, Poznań, Poland
e-mail: witek@amu.edu.pl

T. Żuk
Polish Geological Institute – National Research Institute, Warsaw, Poland

Geoscience FreeD - Open Geomodelling, Warsaw, Poland

T. Radziejewska et al. (eds.), *The Natural Environment of the Mid-Atlantic Ridge*, GeoPlanet: Earth and Planetary Sciences,
https://doi.org/10.1007/978-3-031-51865-2_3

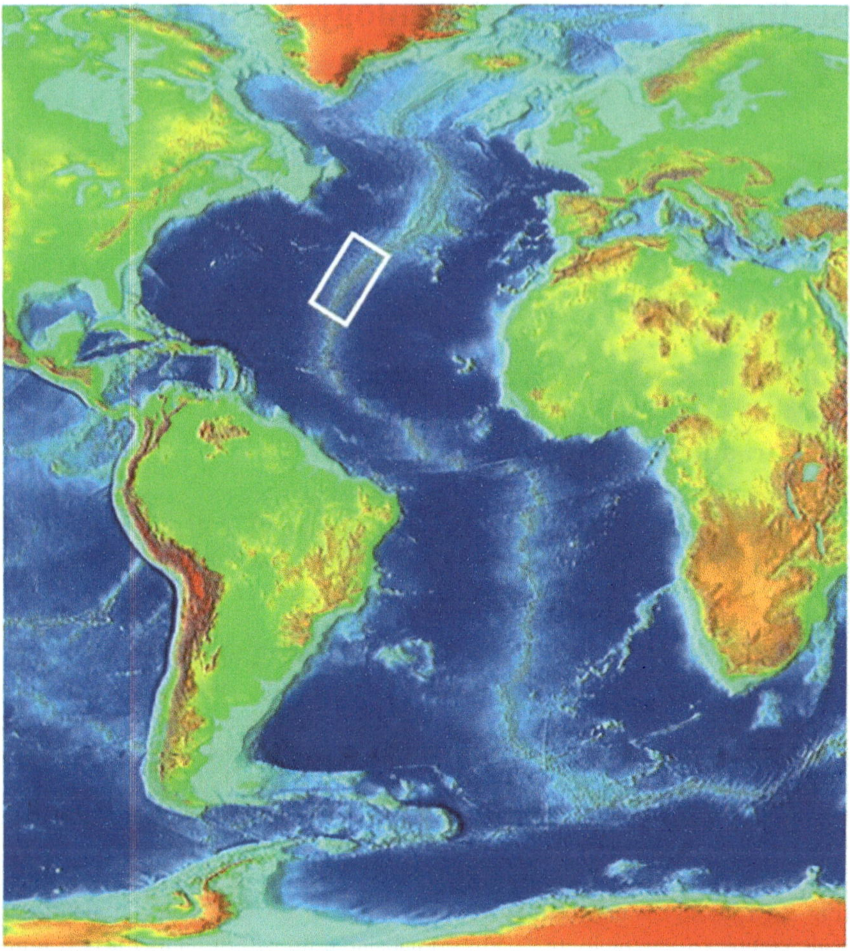

Fig. 1 The general view and bathymetry of the Mid-Atlantic Ridge (MAR) (the approximated area of Polish exploration zone is marked with a white rectangle) (source: http://www.ngdc.noaa.gov/mgg/image/2minrelief.html)

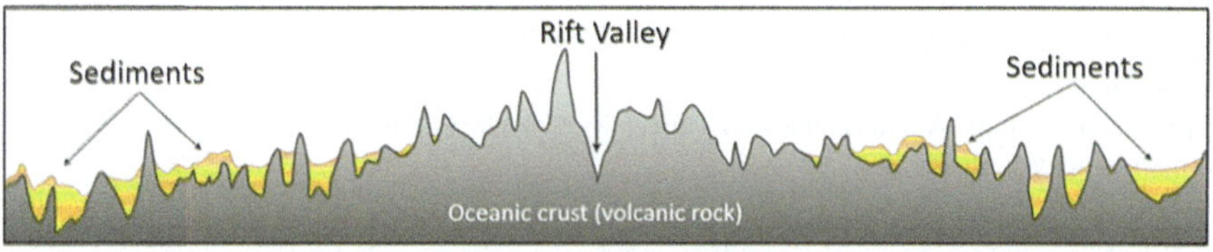

Fig. 2 A cross-section of a typical mid-ocean ridge. Note the generally decreasing elevation from the ridge crest outwards, the rift valley in the axial part of the ridge, rugged topography and the sediment distribution. The sediment cover is thin or absent near the ridge, but it gets generally thicker on either side, as the age of the oceanic crust increases with the distance from the ridge. The cross-section is approximately 50 km long, and has a 10× vertical exaggeration (from Earle 2020)

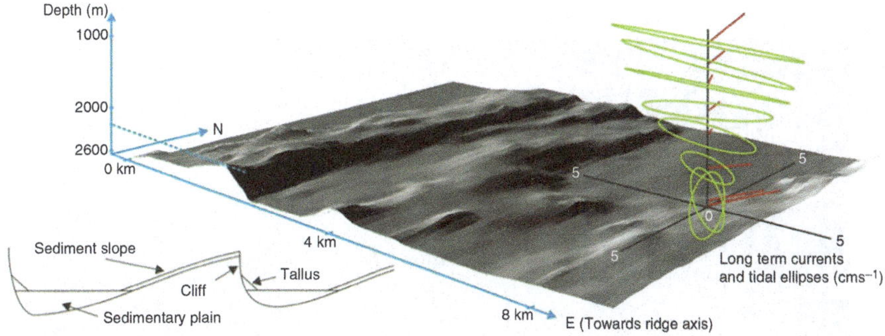

Fig. 3 Seafloor topography and water flow on the flanks of the MAR (around 52° N). The figure presents a 3D projection from swath bathymetry, vertical and horizontal scales are the same; location of the mooring and flow at different heights above the sea floor. Red vectors—long term mean velocity; and a cross-section of a flank of the ridge showing the relationship between flat plains, gentle slopes and steep cliffs. The steep cliffs face towards the ridge axis. (Source: Priede et al. 2013). See Fig. 4 for typical surface deposits along the cross section

Fig. 4 Seafloor images of the Mid-Atlantic Ridge (MAR) (the area identical to that in Fig. 3). Optical ROV images of the seafloor in flat (<5°), gentle-sloping (5°–30°) and steep (>30°) terrains on the MAR. Red or green laser spots are 10 cm apart. (Source: Priede et al. 2013)

pelagic carbonate ooze; Fig. 4) renders—with time and distance from the ridge axis—the seafloor flatter (Figs. 2 and 3) (e.g., Mitchell et al. 1998; Priede et al. 2013).

The first mid-ocean ridge discovered was the Mid-Atlantic Ridge (MAR) (Fig. 1) which was documented already during the famous HMS Challenger expedition in 1872–1876. Later, the MAR was the major case study for the development of the plate tectonic theory, while currently it is studied for various basic science goals as well as for practical reasons. The latter include research on potential environmental impacts of seafloor massive sulphide (SMS) mining.

The Polish exploration area is located within the MAR, in the central part of the Atlantic Ocean (Figs. 1 and 5), and the objective of the present chapter is to summarise the published data on topography, geology, sedimentary processes and sediments in the MAR, with a particular reference to the Polish exploration contract area (ca. 26°–32°N).

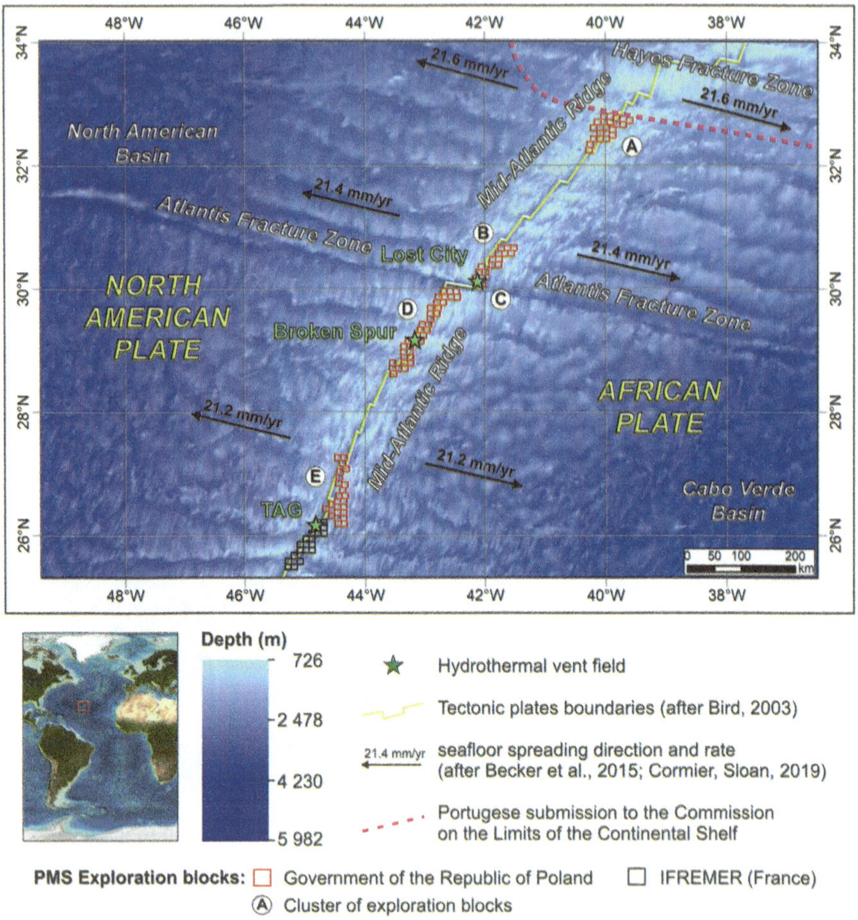

Fig. 5 Location of the Polish polymetallic sulphides (PMS) exploration areas; bathymetric data: GEBCO Compilation Group (2020); maritime boundaries: Flanders Marine Institute (2019)

2 Location, Morphology and Bedrock Geology

The dominant feature of the seafloor morphology in the area is the MAR (Figs. 1, 5, and 6). In this part of the Atlantic Ocean, the MAR trends SW-NE, its relative height being up to 3000 m. From the west, it is flanked by the North American Basin, the Cabo Verde Basin flanking it from the east. The MAR is divided into segments located between transform faults/fracture zones. In the Polish exploration contract area and its immediate vicinity, there are two large fracture zones perpendicular to the ridge: the Atlantis Fracture Zone and the Hayes Fracture Zone (Fig. 5). These fracture zones are narrow valleys with steep walls over 2000 m high. The MAR axis features a rift valley, varying in width between 6000 m in the Atlantis Fracture Zone area to 13,000 m in the southern part of the area. The average relative depth is about 1700 m. The greatest depth of the rift valley (ca. 4500 m) is located in the Atlantis Fracture Zone.

From the tectonic point of view, the area of interest stretches along the divergent boundary of the North American and the African lithospheric plates (Bird 2003) (Figs. 1 and 5). The zones located between transform faults/fracture zones were formed by strike-slip faults cutting through the lithospheric plates and laterally displacing sections of the mid-ocean ridge, sometimes by several hundreds of kilometres. Similarly to faults along the mid-ocean ridge, transform fault zones also act as conduits for fluid migration, which drives hydrothermal processes of the ocean floor. In contrast to the mid-ocean ridge, however, the transform fracture zones usually are free of volcanism, thus the upper flanks of normal faults often expose mafic rocks of the lower lithospheric crust and ultramafic rocks of the upper mantle (Wilson 1965; Batista et al. 2019).

In the area of concern, the spreading rates are relatively low, between 21.2 mm/year in the southern part to 21.6 mm/year in the northern part (Fig. 5) (Becker et al. 2015; Cormier and Sloan 2019). Based on the accretion mode, symmetry of spreading, and type of hosted rocks, two types of slow-spreading ridges have been distinguished: symmetrical accretion with basalts and asymmetrical accretion with gabbro-peridotites (Fig. 6) (McCaig et al. 2007; Escartín et al. 2008; Fouquet et al. 2010). These two geological settings correspond to two types of segments of the slow-spreading MAR: magmatic, dominated by volcanic processes, and tectonic, where faulting processes predominate (German et al. 2016).

The oceanic lithosphere in the discussed MAR area is younger than 10 million years, and the age increases with distance from the ridge axis (Müller et al. 2016)

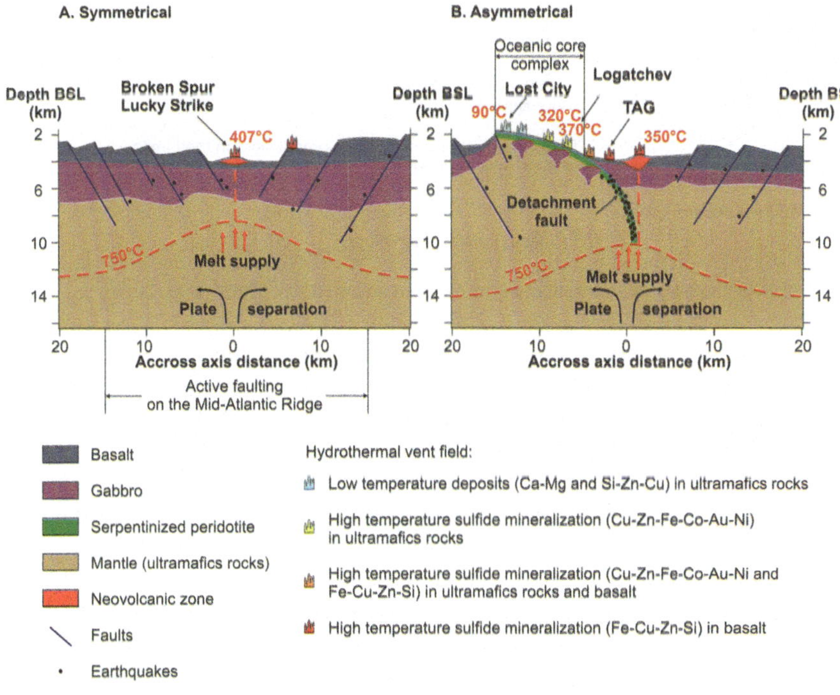

Fig. 6 Two types of slow-spreading ridges (McCaig et al. 2007; Escartín et al. 2008; Fouquet et al. 2010, modified). BSL, below the sea level. Locations are provided in Fig. 5

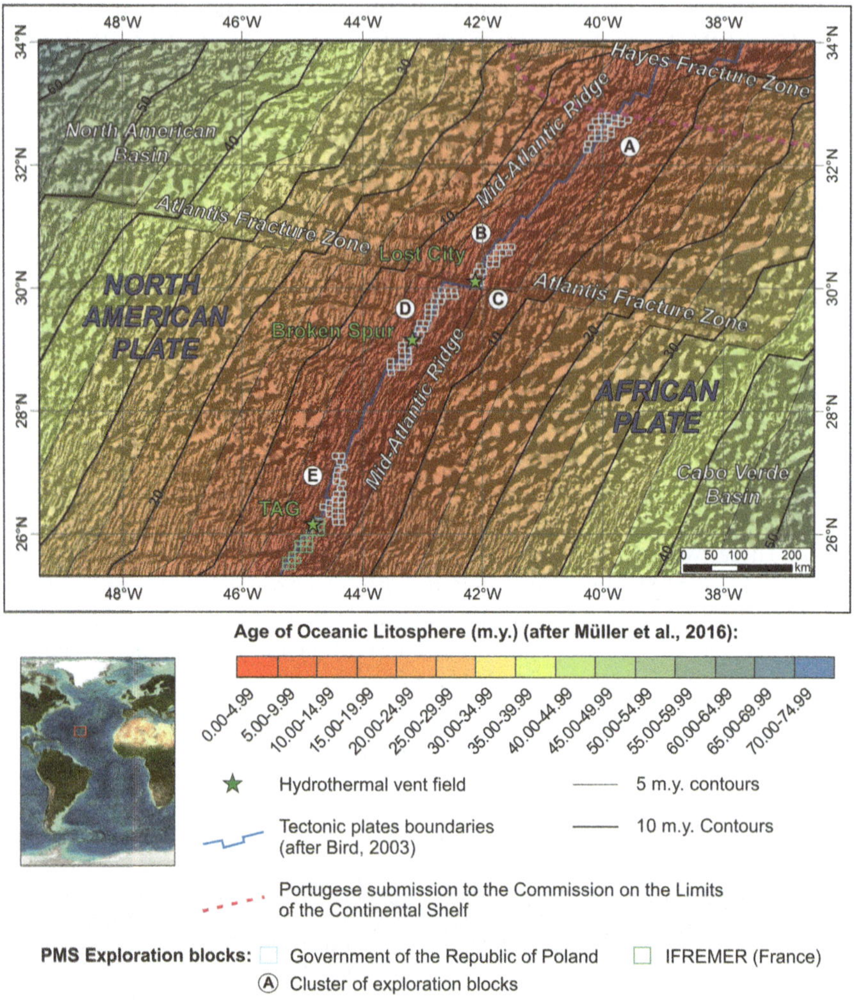

Fig. 7 Age of the oceanic lithosphere in the area of concern (after Müller et al. 2016)

(Fig. 7). Thus, in the area of the Polish exploration blocks, the oceanic lithosphere is younger than 3.3 million years. The seafloor is covered mainly by calcareous ooze (Dutkiewicz et al. 2015) (Fig. 8), as described in more detail below.

The two spectacular features of the area in question are hydrothermal vent fields: Broken Spur and Lost City (Fig. 5). The Broken Spur hydrothermal vent field was discovered in 1993 (Murton 1993; Murton and Van Dover 1993). It is located at 29°10′N on the MAR (Fig. 5). The field is associated with symmetrical seafloor spreading, and is characterized by high-temperature sulphide mineralization (Fig. 6) (McCaig et al. 2007; Escartín et al. 2008; Fouquet et al. 2010). The field covers an area of 150 m from east to west and 100 m from north to south. The Broken Spur includes two parts with black smoker vents and relicts of sulphide/anhydrite

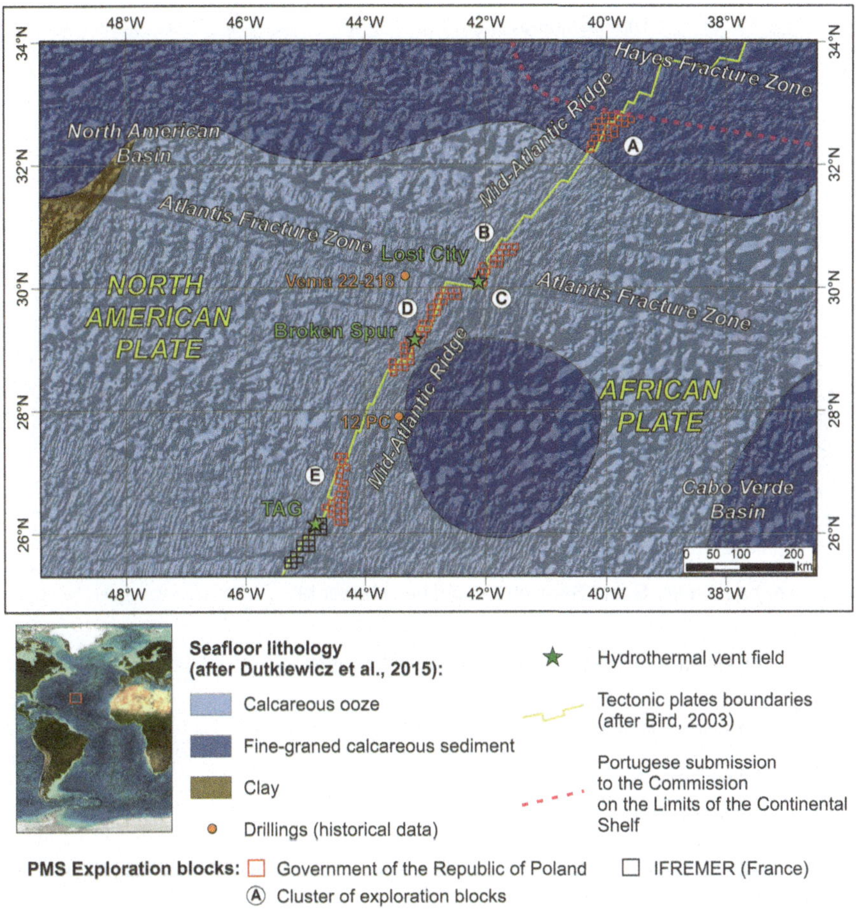

Fig. 8 Seafloor sediments in the area adjacent to the Polish PMS exploration blocks (after Dutkiewicz et al. 2015)

platforms and inactive mounds: the Eastern Valley and the Western Plateau (Fig. 9) (Nesbitt 1995).

The Lost City hydrothermal vent field (Figs. 5, 6, and 10) was discovered in 2000 (Kelley et al. 2001). The field is associated with asymmetrical seafloor spreading, and is characterized by accumulation of low-temperature deposits (Fig. 6) (McCaig et al. 2007; Escartín et al. 2008; Fouquet et al. 2010). It is located within the Atlantis Massif, a well-mapped and -studied oceanic core complex (OCC) at ca. 30°N, on the elevation margin between the Atlantis Transform Fault and the MAR (Blackman et al. 2002; Kelley et al. 2005; Karson et al. 2006; Blackman et al. 2008, 2009; Kelley and Shank 2010) (Figs. 5 and 10). The field hosts carbonate towers rising 60 m above the seafloor (Kelley et al. 2001, 2005; Proskurowski et al. 2008; Lang et al. 2010).

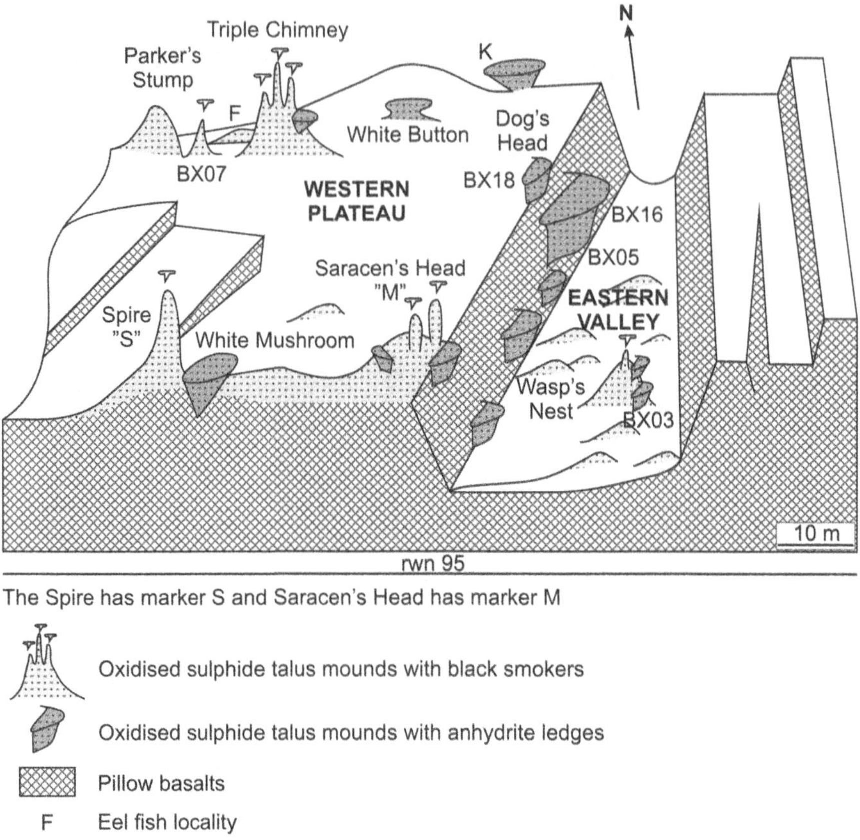

Fig. 9 Morphology of the seafloor around Broken Spur (Nesbitt 1995). For the field location and additional data see Figs. 5 and 6

The southern exposure of the Atlantis Massif rises to 3800 m from the base of the Atlantis Transform Fault, culminating in a cliff face about 200 m high (Fig. 10). The cliff rises to a water depth of 740 m below the sea level. It is deeply embayed due to mass wasting and faulting, which exposes massive serpentinite with gabbroic rocks. The wall terminates in a 50–100 m thick, variably deformed zone that defines the main Atlantis Massif Detachment Shear Zone at the summit of the massif. Over the summit of the Atlantis Massif, there lies unconformably a polymict breccia (rounded clasts of serpentinite, gabbro, talc-tremolite nodules, and subangular weathered basalt in a micritic carbonate matrix) and carbonate caprock (pelagic fossil-rich micritic carbonate and hydrothermal carbonate). These deposits are up to 2 m thick (Kelley et al. 2001, 2005; Früh-Green et al. 2003; Karson et al. 2006; Denny et al. 2015).

The main part of the field is at a water depth of about 740 m (Fig. 10). The core of the field is dominated by the Poseidon complex comprised of four adjoining massive edifices. Discrete chimneys, and chimney arrays, extend from the Poseidon complex northward and to the east (Kelley et al. 2005; Denny et al. 2015). The main tower rises 60 m above the seafloor and tapers to a summit 15 m in diameter. The

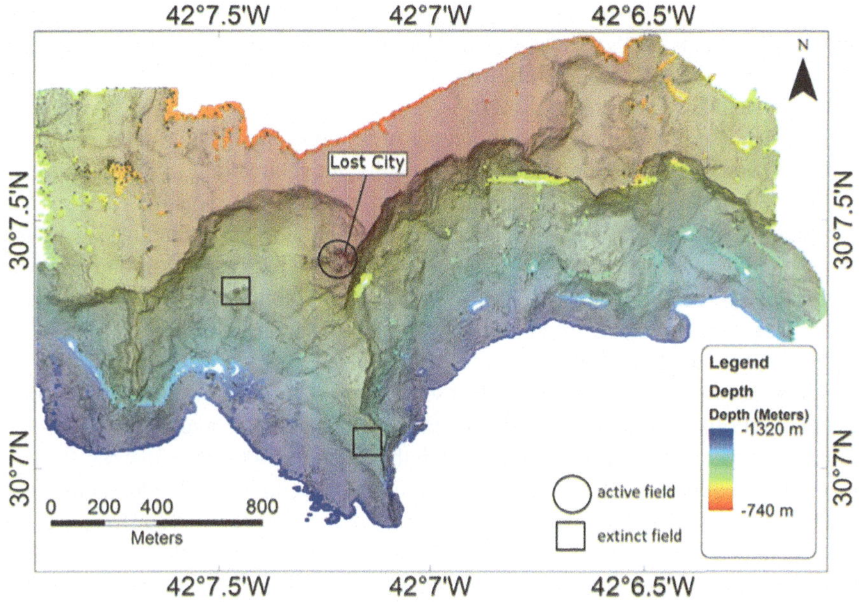

Fig. 10 Bathymetry of the Atlantis Massif summit with the Lost City hydrothermal vent field (Denny et al. 2015)

bulk of the main tower is composed of indurated carbonate heavily crosscut with more recent carbonate veins. The summit of the tower hosts two carbonate structures venting fluids at up to 88 °C (Ludwig et al. 2006). Age-dating of 67 carbonate deposits within the central Lost City hydrothermal field yielded the ages ranging from modern to 120,000 years old (Ludwig et al. 2011). In addition, two extinct hydrothermal fields were discovered within the area (Fig. 10).

3 Deep-Sea Sediments

The sediment distribution on the MAR has been of interest already since the 1960s, as the sediment cover thickness was used as an indicator of the seafloor spreading rate and an argument on the nature of this process (continuous vs. stepwise) (e.g., Ewing and Ewing 1967). However, although the sediment thickness is generally expected to increase systematically with distance from the spreading axis, as would be expected from the increasing seafloor age (Figs. 2 and 7), it is not always the case (see Fig. 3). For instance, Mitchell et al. (1998) applied a deep-tow sediment profiler system in the ridge area around 29°N, and found the sediments on the young (<2.5 million years) seafloor, at a distance up to about 40 km west and east from the spreading axis, respectively, to reveal no consistent thickening with distance from the axis. They have documented thicker deposits in depressions, with sediment thickness on the order of 10–15 m, in some cases >25 m. Many of the deposits appeared to drape the basement topography and occurred also on promontories, so the sediment was probably mainly deposited from suspension rather than from turbidity currents or other gravity flows.

Enhanced sedimentation rates in deeper areas could partly be due to redeposition, as the MAR is well-known as a seismically active area (e.g., Fig. 6). Mitchell et al. (1998) estimated also the sediment accumulation rate over the MAR. They computed sediment accumulation rates by dividing the average thicknesses by the corresponding basement ages estimated from magnetic field reversals, and came up with values on the order of 5–24 mm/kyr. This range is similar to the sediment accumulation rate estimates from elsewhere along the MAR, for instance along 37°N (Marks 1981; Mitchell 1995).

There are only several published observational data on the seafloor sediments within the Polish exploration blocks on the MAR. In general, since the area is shallower than the carbonate compensation depth (CCD; ca. 5000 m below the sea surface), the seafloor surface sediments are composed mainly of calcareous ooze (Dutkiewicz et al. 2015) (Fig. 8). In the area of interest, as few as four longer cores were obtained (the IODP site in the Lost City, Vema 22-218, 12PC, and TAG; see Fig. 8 for locations). In the IODP drilling within the Lost City, the bottom of the calcareous ooze was reached at 2 m below the seafloor (Denny et al. 2015). The rock samples obtained from dredges and IODP drillings (https://maps.ngdc.noaa.gov/viewers/marine_geology) implied that the calcareous ooze was underlined by igneous rocks (porphyritic olivine basalt, feldspar-rich basalt, glassy basalt) and serpentine with relics of pyroxene and olivine (Denny et al. 2015).

The thickest cover of at least 5.85 m (the bottom of the sedimentary cover was not reached during the coring) was found in the Atlantis Fracture Zone, 115 km west of the Lost City hydrothermal vent field in core Vema 22-218 (https://maps.ngdc.noaa.gov/viewers/marine_geology) (Fig. 8). The sediment in that core was grayish-orange to moderate yellowish-brown, and the vertical sedimentary profile was composed of intercalated layers of foraminiferal chalk, marl and ooze. The coarse fraction was made up of planktonic and benthic foraminifera as well as rare pteropods. Moreover, infrequent manganese micronodules and andesite-basaltic rock fragments were encountered.

Core No. 12PC, taken on the MAR in an area between the D and E exploration cluster blocks (https://maps.ngdc.noaa.gov/viewers/marine_geology) (Fig. 8), showed 5.17 m of calcareous ooze. The bottom of the series was not reached. The coarse fraction was made up by foraminifera. The ooze was light-yellow-brown to very pale brown.

In the southern part of the studied area, in the axial part of the MAR, there is the Trans-Atlantic Geotraverse (TAG) hydrothermal field at about 26°N. Metz et al. (1988) have collected and investigated a 120 cm-long core from metalliferous sediment, covering ca. 14,000 years of hydrothermal depositional history in that area. During the periods of high hydrothermal activity, the sediments were composed almost entirely of hydrothermally derived elements (Fe, Cu, Zn, Mn, V, Pb, Cd, Hg) with less than 3% of carbonates. During periods of decreased hydrothermal activity, the sediments were composed in over 60% of carbonates. The calculated sediment accumulation rates varied in that core from 1 to >30 g/cm^2/kyr throughout the core.

4 Sediment Sources and Sedimentation Processes

The scarcity of data on sediments and sedimentary processes within the area of concern does not allow for any comprehensive overview of the sedimentary regime over the MAR between 26° and 34°N. However, the available information on similar sediment accumulation rates in this part of the MAR (Mitchell et al. 1998) and in the sectors nearby (e.g., Marks 1981) as well as similar general properties of the sediments along most of the northern MAR (Dutkiewicz et al. 2015; Khusid et al. 2018; Fig. 8) suggest similarity of some major sediment sources and the sedimentation processes.

With regard to the MAR sediment sources, an interesting study based on mineralogical and geochemical properties of mid-oceanic ridge sediments (for the section between 45° and 65 N) was conducted by Grousset and Chesselet (1986). They focussed on the sediment mineral fraction (so the analysis excluded the dominant biogenic carbonates). Using a number of tracers, they identified the North America mainland (Fig. 1) as the main source of the sediment supplied by surface currents, turbidity currents and aeolian transport, Iceland being the second major source, with the material transported by turbidity currents and the geostrophic current flow. The MAR itself was found to supply only a minor amount of the sedimentary material. The MAR section of interest for this report is more southward. It is located further away from the North America mainland and Iceland, while the Azores are much closer, so the relative contribution of the particular sources may be different.

In terms of the acting processes, it seems that sedimentation from the pelagic zone is the dominant mode of deposition (Mitchell et al. 1998). However, the existing data suggest also a likely sediment redistribution. The latter is facilitated by seismic activity mobilising sediments on the slopes (e.g., Figs. 3 and 4), regular near-bottom currents (e.g., Priede et al. 2013; Fig. 3), or benthic storms (Kerr 1980). The mineral fraction of the MAR may be easily mobilised because of its very fine fraction, often in the fine silt range (Alt et al. 2019). This sediment fraction renders itself easier to resuspend than the smaller clay fraction, as cohesion forces are of lesser importance. The resuspension susceptibility of the sediment must be considered as an important environmental issue, particularly in the regions where seafloor massive sulphides (SMS; also referred to as polymetallic massive sulphides, PMS) are intended for mining (e.g., Dunn et al. 2018). It is exceptionally important in the case of sediments sourced from hydrothermal vents, which contain very high concentrations of potentially toxic metals and metalloids (e.g., Metz et al. 1988; Dias et al. 2008).

5 Conclusions

In terms of the sea-floor morphology and some of the geological problems, the Polish exploration contract area on the MAR is moderately known. However, the current knowledge of sediments and sedimentary processes is very limited. The sedimentary cover is from few to over 20 m thick, thicker in depressions, composed mainly of carbonate ooze, which in the neighbourhood of hydrothermal vents may be extremely enriched in metals and metalloids. Any future activity in the region will need to include sediment mapping, to consider analysis of modern sedimentation processes and sediment sources, and will have to assess a potential impact of metal-enriched sediment resuspension on the benthic ecosystems.

Authors' Contribution P.S. with input from T.Ż and A.K.-R. wrote the original version of sections 2 and partly 3 (including figures 5–10), W.S. wrote the remaining sections 1, 4, 5, and most of section 3, and has revised section 2.

References

Alt CHS, Kremenetskaia A, Gebruk AV, Gooday AJ, Jones DOB (2019) Bathyal benthic megafauna from the Mid-Atlantic Ridge in the region of the Charlie-Gibbs Fracture Zone based on remotely operated vehicle observations. Deep Sea Res Part I Oceanogr Res Pap 145:1–12. https://doi.org/10.1016/j.dsr.2018.12.006

Batista L, Gasperini L, Burwicz EB, Neres M, Nuzzo M (2019) Marine transform faults and fracture zones: a joint perspective integrating seismicity, fluid flow and life. Front Earth Sci 7:1–29. https://doi.org/10.3389/feart.2019.00039

Becker TW, Schaeffer AJ, Lebedev S, Conrad CP (2015) Toward a generalized plate motion reference frame. Geophys Res Lett 42:3188–3196. https://doi.org/10.1002/2015GL06369

Bird P (2003) An updated digital model of plate boundaries. Geochem Geophys Geosyst 4:1027. https://doi.org/10.1029/2001GC000252

Blackman DK, Karson JA, Kelley DS, Cann JR, Früh-Green GL, Gee JS, Hurst SD, John BE, Morgan J, Nooner SL, Kent Ross D, Schroeder TJ, Williams EA (2002) Geology of the Atlantis Massif (Mid-Atlantic Ridge, 30°N): implications for the evolution of an ultramafic oceanic core complex. Mar Geophys Res 23:443–469. https://doi.org/10.1023/B:MARI.0000018232.14085.75

Blackman DK, Karner GD, Searle RC (2008) Three-dimensional structure of oceanic core complexes: effects on gravity signature and ridge flank morphology, Mid-Atlantic Ridge, 30°N. Geochem Geophys Geosyst 9(6):Q06007. https://doi.org/10.1029/2008GC001951

Blackman DK, Canales JP, Harding A (2009) Geophysical signatures of oceanic core complexes. Geophys J Int 178:593–613. https://doi.org/10.1111/j.1365-246X.2009.04184.x

Cormier MH, Sloan H (2019) Distinctive seafloor fabric produced near western versus eastern ridge transform intersections of the northern mid-Atlantic ridge: possible influence of ridge migration. Geochem Geophys Geosyst 20(4):1734–1755. https://doi.org/10.1029/2018GC008101

Denny AR, Kelley DS, Früh-Green GL (2015) Geologic evolution of the Lost City hydrothermal field. Geochem Geophys Geosyst 17:375–394. https://doi.org/10.1002/2015GC005869

Dias AS, Millis RA, Taylor RN, Ferreira P, Barriga FJAS (2008) Geochemistry of a sediment pushcore from the Lucky Strike hydrothermal field, Mid-Atlantic Ridge. Chem Geol 247:339–351. https://doi.org/10.1016/j.chemgeo.2007.10.015

Dunn DC, Van Dover SL, Etter RJ, Smith CR, Levin LA, Morato T, Colaço A, Dale AC, Gebruk AV, Gjerde KM, Halpin PN, Howell KL, Johnson D, Perez JAA, Ribeiro MC, Stuckas H, Weaver P, SEMPIA Workshop Participants (2018) A strategy for the conservation of biodiversity on mid-ocean ridges from deep-sea mining. Sci Adv 4:eaar4313. https://doi.org/10.1126/sciadv.aar4313

Dutkiewicz A, Müller RD, O'Callaghan S, Jóson H (2015) Census of seafloor sediments in the world's ocean. Geology 43:795–798. https://doi.org/10.1130/G36883.1

Earle S (2020) Physical geology. 2nd edn. Thompson Rivers University, Kamloops

Escartín J, Smith DK, Cann J, Schouten H, Langmuir CH, Escrig S (2008) Central role of detachment faults in accretion of slow-spreading oceanic lithosphere. Nature 455:790–794. https://doi.org/10.1038/nature07333

Ewing J, Ewing M (1967) Sediment distribution on the mid-ocean ridges with respect to spreading of the sea floor. Science 156:1590–1592. https://doi.org/10.1126/science.156.3782.1590

Flanders Marine Institute (2019) Maritime boundaries geodatabase, version 11. http://www.marineregions.org/. https://doi.org/10.14284/382

Fouquet Y, Cambon P, Etoubleau J, Charlou JL, Ondreas H, Barriga FJAS, Cherkashov G, Semkova T, Poroshina I, Bohn M, Donval JP, Henry K, Murphy P, Rouxel O (2010) Geodiversity of hydrothermal processes along the Mid-Atlantic Ridge and ultramafic hosted mineralization: a new type of oceanic Cu-Zn-Co-Au volcanogenic massive sulfide deposit. Geophys Monogr Ser 188:321–367. https://doi.org/10.1029/2008GM000746

Früh-Green GL, Kelley DS, Bernasconi SM, Karson JA, Ludwig KA, Butterfield DA, Boschi C, Proskurowski G (2003) 30,000 years of hydrothermal activity at the Lost City vent field. Science 301:495–498. https://doi.org/10.1126/science.1085582

GEBCO Compilation Group (2020). GEBCO 2020 Grid. https://doi.org/10.5285/a29c5465-b138-234d-e053-6c86abc040b9

German C, Petersen S, Hannington MD (2016) Hydrothermal exploration of midocean ridges: where might the largest sulfide deposits occur? Chem Geol 420:114–126. https://doi.org/10.1016/j.chemgeo.2015.11.006

Grousset FE, Chesselet R (1986) The Holocene sedimentary regime in the northern Mid-Atlantic Ridge region. Earth Planet Sci Lett 78:271–287. https://doi.org/10.1016/0012-821X(86)90067-1

https://maps.ngdc.noaa.gov/viewers/marine_geology - drillings profiles and drillings location data (version 2.7.0, last updated 5/17/2019)

Karson JA, Früh-Green GL, Kelley DS, Williams EA, Yoerger DR, Jakuba M (2006) Detachment shear zone of the Atlantis Massif core complex, Mid-Atlantic Ridge, 30°N. Geochem Geophys Geosyst 7:Q06016. https://doi.org/10.1029/2005GC001109

Kelley DS, Shank TM (2010) Hydrothermal systems: a decade of discovery in slow spreading environments. In: Diversity of hydrothermal systems on slow spreading ocean ridges, vol 188. Wiley, pp 369–407. https://doi.org/10.1029/2010GM000945

Kelley DS, Karson JA, Blackamn DK, Früh-Green GL, Butterfield DA, Lilley MD, Olson EJ, Schrenk MO, Roe KK, Lebon GT, Rivizzigno P, the AT3-60 Shipboard Party (2001) An off-axis hydrothermal vent field near the Mid-Atlantic Ridge at 30°N. Nature 412:145–149. https://doi.org/10.1038/35084000

Kelley DS, Karson JA, Früh-Green GL, Yoerger DR, Shank TM, Butterfield DA, Hayes JM, Schrenk MO, Olson EJ, Proskurowski G, Jakuba M, Bradley A, Larson B, Ludwig K, Glickson D, Buckman K, Bradley AS, Brazelton WJ, Roe K, Elend MJ, Delacour A, Bernasconi SM, Lilley MD, Baross JA, Summons RE, Sylva SP (2005) A serpentinite-hosted ecosystem: the Lost City hydrothermal field. Science 307:1428–1434. https://doi.org/10.1126/science.1102556

Kerr RA (1980) A new kind of storm beneath the sea. Science 208:484–486. https://doi.org/10.1126/science.208.4443.484

Khusid TA, Os'kina NS, Lukashina NP, Gablina IF, Libina NV, Matul G (2018) Benthic and plankton foraminifers in hydrothermally active zones of the Mid-Atlantic Ridge (MAR). Stratigr Geol Correl 26:109–120. https://doi.org/10.1134/S0869593817060041

Lang SQ, Butterfield DA, Schulte M, Kelley DS, Lilley MD (2010) Elevated concentrations of formate, acetate and dissolved organic carbon found at the Lost City hydrothermal field. Geochim Cosmochim Acta 74(3):941–952. https://doi.org/10.1016/j.gca.2009.10.045

Ludwig KA, Kelley DS, Butterfield DA, Nelson BK, Früh-Green GL (2006) Formation and evolution of carbonate chimneys at the Lost City hydrothermal field. Geochim Cosmochim Acta 70:3625–3645. https://doi.org/10.1016/j.gca.2006.04.016

Ludwig KA, Shen CC, Kelley DS, Cheng H, Edwards RL (2011) U-Th systematics and ^{230}Th ages of carbonate chimneys at the Lost City hydrothermal field. Geochim Cosmochim Acta 75(7):1869–1888. https://doi.org/10.1016/j.gca.2011.01.008

Marks NS (1981) Sedimentation on new ocean crust: the Mid-Atlantic Ridge at 37°N. Mar Geol 43:65–82. https://doi.org/10.1016/0025-3227(81)90129-8

McCaig AM, Cliff RA, Escartín J, Fallick AE, MacLeod CJ (2007) Oceanic detachment faults focus very large volumes of black smoker fluids. Geology 35(10):935–938. https://doi.org/10.1130/G23657A.1

Metz S, Trefry JH, Nelsen TA (1988) History and geochemistry of a metalliferous sediment core from the Mid-Atlantic Ridge at 26°N. Geochim Cosmochim Acta 52:2369–2318

Mitchell NC (1995) Diffusion transport model for pelagic sediments on the Mid-Atlantic Ridge. J Geophys Res 100:19991–20009

Mitchell NC, Allerton S, Escartin J (1998) Sedimentation on young ocean floor at the Mid-Atlantic Ridge, 29°N. Mar Geol 148:1–8

Müller RD, Seton M, Zahirovic S, Williams SE, Matthews KJ, Wright NM, Shephard GE, Maloney KT, Barnett-Moore N, Hosseinpour M, Bower DJ, Cannon J (2016) Ocean basin evolution and global-scale plate reorganization events since Pangea breakup. Annu Rev Earth Planet Sci 44:107–138. https://doi.org/10.1146/annurev-earth-060115-012211

Murton BJ (1993) Geological and geochemical investigations between 27°N and 30°N of the Kane-Atlantis segment: the Mid-Atlantic Ridge. Institute of Oceanographic Sciences Deacon Laboratory, Cruise Report No. 236

Murton BJ, Van Dover CL (1993) ALVIN dives on the Broken Spur hydrothermal vent field at 29°10′N on the Mid-Atlantic Ridge. BRIDGE Newsl 5:11–14

Nesbitt RW (1995) The geology of the Broken Spur hydrothermal vent site; a new look at an old field. BRIDGE Newsl 8:30–34

Priede IG, Bergstad OA, Miller PI, Vecchione M, Gebruk A, Falkenhaug T, Billett DS, Craig J, Dale AC, Shields MA, Tilstone GH, Sutton TT, Gooday AJ, Inall ME, Jones DO, Martinez-Vicente V, Menezes GM, Niedzielski T, Sigurðsson Þ, Rothe N, Rogacheva A, Alt CH, Brand T, Abell R, Brierley AS, Cousins NJ, Crockard D, Hoelzel AR, Høines Å, Letessier TB, Read JF, Shimmield T, Cox MJ, Galbraith JK, Gordon JD, Horton T, Neat F, Lorance P (2013) Does presence of a mid-ocean ridge enhance biomass and biodiversity? PLoS One 8:e61550. https://doi.org/10.1371/journal.pone.0061550

Proskurowski G, Lilley MD, Seewald JS, Früh-Green GL, Olson EJ, Lupton JE, Sylva SP, Kelley DS (2008) Abiogenic hydrocarbon production at Lost City hydrothermal field. Science 319:604–607. https://doi.org/10.1126/science.1151194

Wilson JT (1965) A new class of faults and their bearing on continental drift. Nature 207:343–347. https://doi.org/10.1038/207343a0

Part II
Biology and Ecology

Pelagic Environment and Its Biocoenosis

Katarzyna Błachowiak-Samołyk, Tomasz B. Linkowski, Lech Stempniewicz, Michał Tomczak, Jan Marcin Węsławski, and Bartosz Kotrys

Deep-sea pelagic habitats include the dark ocean (>200 m depth) beneath the epipelagic (0–200 m depth) layer as well as the mesopelagic (200–1000 m) and bathypelagic (water column >1000 m) zones (Figs. 1 and 2). Except for limited areas of hydrothermal vents and seeps, life patterns in deep-sea pelagic habitats along the Mid-Atlantic Ridge (MAR) depend on biological production exported from the sunlit epipelagic layers (Vecchione et al. 2010a, b). In this section, we will describe individual components of the pelagic environment in the MAR region (Fig. 3), following the size categories of biota as presented in Fig. 4.

K. Błachowiak-Samołyk (✉) · J. M. Węsławski
Institute of Oceanology PAN, Sopot, Poland
e-mail: kasiab@iopan.gda.pl; weslaw@iopan.gda.pl

T. B. Linkowski
Sea Fisheris Institute - National Research Institute, Gdynia, Poland
e-mail: tlinkowski@mir.gdynia.pl

L. Stempniewicz
Faculty of Biology, University of Gdansk, Gdańsk, Poland
e-mail: lech.stempniewicz@ug.edu.pl

M. Tomczak · B. Kotrys
Polish Geological Institute – National Research Institute, Warsaw, Poland
e-mail: mtomc@pgi.gov.pl; bkotr@pgi.gov.pl

T. Radziejewska et al. (eds.), *The Natural Environment of the Mid-Atlantic Ridge*, GeoPlanet: Earth and Planetary Sciences,
https://doi.org/10.1007/978-3-031-51865-2_4

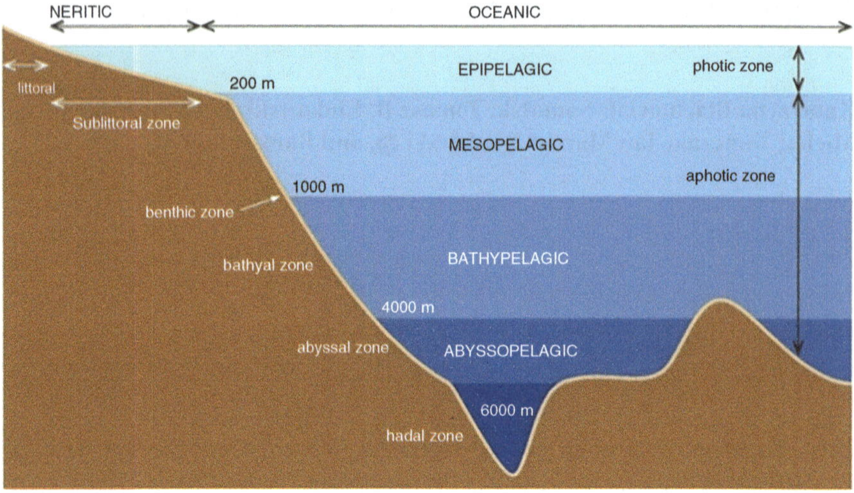

Fig. 1 Division of the oceanic domain into major ecological zones (compilation from various sources)

Fig. 2 Division of the oceanic system into ecological zones and the representative organisms (compilation from various sources)

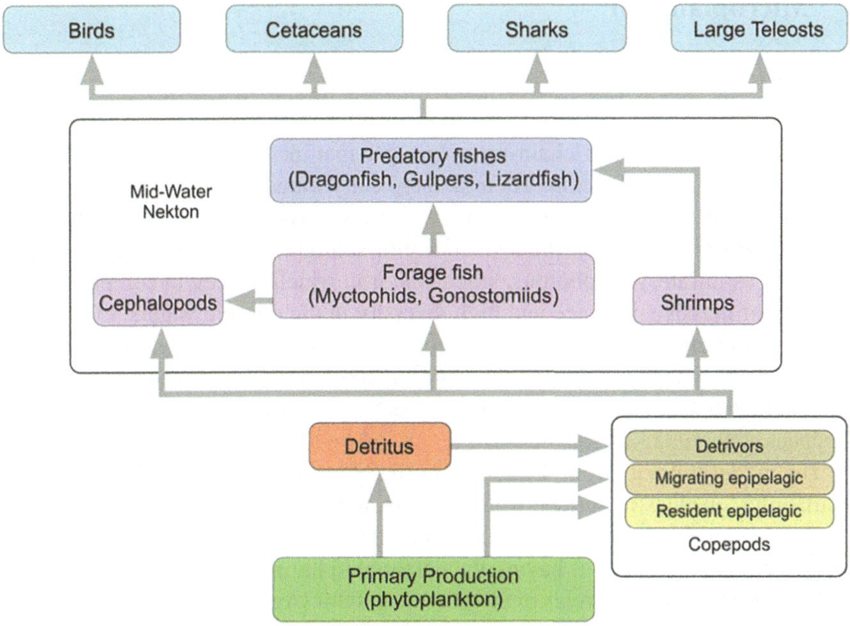

Fig. 3 A schematic diagram of the pelagic trophic web in the mid-Atlantic region (modified after Anderson et al. 2019)

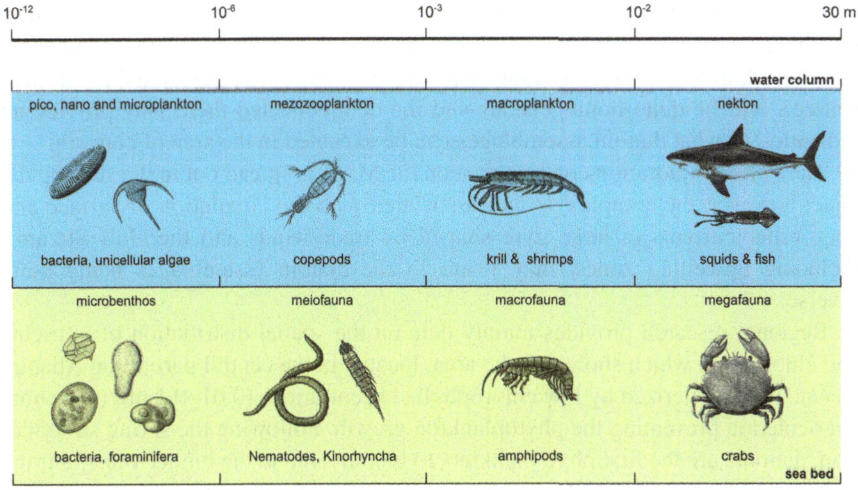

Fig. 4 Main size-based ecological categories of oceanic organisms in the water column and on the seabed, with dominant taxa indicated (compilation from various sources)

1 Microplankton

1.1 Diatoms

Diatoms are unicellular or chain-forming algae that incorporate silicon in the construction of their exoskeleton. Their size typically ranges from 1.5 to 500 μm and the number of diatom species is estimated at over 100,000 (Norton et al. 1996). Diatom growth depends on the availability of sunlight, carbon dioxide, and nutrients such as nitrate, phosphorous, silica, and iron which are key in cell production. Because of massive presence and high diversity in the oceanic waters, diatoms are the most important component of phytoplankton and provide a significant amount of the world's oxygen supply (25%), and the main source of marine primary production (up to 40%), which makes them crucial for ocean biological carbon pump (Tréguer et al. 1995, 2018).

The modern distribution of diatoms in the North Atlantic depends on nutrients availability, changes in surface water salinity, temperature, and circulation pattern resulting from ocean surface conditions and atmospheric forcing. These ecological parameters are important as having the potential to impact the marine food chain as well as biogeochemical cycles or the climate system (Agusti et al. 2015; Allen et al. 2005; Sieracki et al. 1993).

The available literature shows that the research area (central Mid-Atlantic Ridge) is understudied in terms of diatom identification and the taxonomic recognition of species. The available data collected during research conducted in the Polish contract area addressed mainly the zooplankton, and to a lesser extend the radiolarians (e.g. Vereshchaka and Vinogradov 1999). As the Atlantic Ocean surface waters are well mixed, it may be assumed that the axial part of the ocean, far from the sediment sources, will be quite homogeneous and the data collected there may provide an indication of what diatom assemblages can be expected in the area of concern.

The Polish exploration contract area on the MAR is spread out in the subtropical zone; however, the complex system associated with the circulation of surface and deep-water currents (a large gyre shaped by trade-winds and the Gulf Stream), including upwelling zones, may result in the diatom assemblages being quite diverse.

Regional research provides mainly data on the spatial distribution of nutrients and chlorophyll which show that the area, located in the central part of the Atlantic Ocean, is characterized by low chlorophyll concentrations (0.01–0.1 mg/m^3), nutrient depletion preventing the phytoplankton growth. Following the spring stratification, diatoms are the first phytoplankters to bloom. Due to the higher silica, nitrate, and phosphate availability, the northern Atlantic (and the Arctic) waters are characterized by a high abundance of diatoms (Olguin et al. 2006; Barton et al. 2015; Oksman et al. 2019). The highest values (>100 × 10^6 cells/L) have been recorded along the shelf areas, while the highest values in the open Atlantic Ocean have been reported from south of Iceland (10 × 10^6 cells/L) to decrease towards the central part of the North Atlantic to an average of >0.25 × 10^6 cells/L, to increase again around the Equator to about 3–4 × 10^6 cells/L. The central North Atlantic features a lower primary production in the photic zone and a very low diatom abundance in the surface water; consequently, the surface sediment show few diatoms only.

The published data show diatom assemblages in the central North Atlantic to have very low diversity and concentration in the surface water, the concentrations varying between 1 and 0.05 × 10^6 cells/L. The assemblages are dominated by cosmopolitan planktonic diatoms (the genera *Chaetoceros* and *Bacteriastrum* account for 50% of the assemblage), centric and pennate diatoms contributing 40 and 10%, respectively. Within the central North Atlantic, more than 75 diatom species have been identified, a much higher species richness occurring closer to the Equator and towards the coasts (Maynard 1976; Schrader et al. 1993). Some of the common species are listed in Table 1.

From the perspective of the planned exploration, it is necessary to take into account the seasonality, which shows a close relationship between the physical parameters and the diversity of diatom communities. Generally, a weaker stratification, increased water mixing, turbulence, and reduced transparency, while maintaining access to light and nutrients, bring about an increase in diatoms abundance in the central ocean areas of the subtropical zone. A potential exploitation of massive sulphides from the bottom of the ridge, which would deliver suspended particles to the water column because of the plume discharge (a material with toxic properties, but enriched with silica), may result in diatom blooms in the surface waters. The onset of mining may cause an increase in productivity of the photic zone, leading to an increase in the amount of particulate organic matter descending to the bottom (known as the marine snow), which will lead to further changes in the food chain.

Table 1 Common species in the diatom assemblages from central MAR (compilation from different sources)

Temperate water species	Warm water species
Chaetoceros contortus Schütt, 1895	*Chaetoceros danicus* Cleve, 1889
Chaetoceros rostratus Ralfs, 1864	*Coscinodiscus lineatus* Ehrenberg, 1839
Cerataulina pelagica (Cleve) Hendey, 1937	*Coscinodiscus nodulifer* A.W.F. Schmidt, 1878
Detonula pumila (Castracane) Gran, 1900	*Coscinodiscus tuberculatus* Greville, 1861
Coscinodiscus granii L.F. Gough, 1905	*Mastogloia rostrata* (Wallich) Hustedt, 1933
Leptocylindrus danicus Cleve, 1889	*Nitzschia marina* Grunow, 1880
Minutocellus polymorphus (Hargraves & Guillard) Hasle, Stosch & Syvertsen, 1983	*Thalassionema nitzschioides* (Grunow) Mereschkowsky, 1902
Pseudo-nitzschia multiseries (Hasle) Hasle, 1995	*Thalassiosira eccentrica* (Ehrenberg) Cleve, 1904
Rhizosolenia bergonii H. Peragallo, 1892	*Thalassiosira oestrupii* (Ostenfeld) Hasle, 1972
Rhizosolenia styliformis T. Brightwell, 1858	
Thalassiosira decipiens (Grunow) Jørgensen, 1905	
Thalassiothrix longissima Cleve & Grunow, 1880	

As a result, the diatom assemblages may become similar to those in upwelling zones. A similar increase in productivity of the Atlantic surface waters occurs when the ocean surface is fertilised by dust carried by the eeolian transport from the Sahara Desert.

1.2 Radiolaria

Siliceous microorganisms, including radiolarians (Radiolaria), are marine planktonic protozoans which range in size from 20 up to 300 μm and are widely distributed in the oceans. They occur throughout the water column, from near the surface to the deep ocean (Anderson 1983). Their production in the northern hemisphere is highest in July–September and decreases in October–June (Voituriez and Herbland 1977). Under favourable conditions, radiolarian tests preserve very well in the sediment and are an important component of the siliceous ooze (De Wever et al. 2001; Kotrys 2012).

The spatial distribution of living radiolarians reflects primarily the temperature of the surface and subsurface water (Haq and Boersma 1998), but the radiolarian habitat is also influenced by a combination of other environmental factors such as the nutrient availability, dissolved silica content, seasonal productivity regime and salinity (Bjørklund et al. 1998). Therefore, the highest radiolarian abundances are restricted to upwelling and silica sink zones, particularly in the Southern Ocean (de Master 1981).

Dissolved silica in the water column and the fossilized silica content in the sediment decline sharply towards low latitudes of the central part of the Atlantic Ocean, for which reason the radiolarian zooplankton occurrence in the Polish contract area on the Mid-Atlantic Ridge (MAR) is very limited. However, the limitation may also result from the existing knowledge gap. In fact, the radiolarian zooplankton in the subtropical-tropical areas has been poorly studied, compared to other World Ocean regions. Boltovskoy et al. (1996) conducted one of the very few studies addressing radiolarians of the water column in the vicinity of the MAR (20°55.3′N; 19°44.5′W). Based on that study, its seems likely that some of the radiolarian species listed in Table 2 occur in the Polish contract area.

Vereshchaka and Vinogradov (1999) who performed visual censuses of the zooplankton over the Broken Spur hydrothermal field (in the Polish contract area) reported the two zooplankton biomass aggregations they observed along the vertical profiles over the vents to be dominated by gelatinous zooplankton and by radiolarians. For obvious reasons (no actual sampling), they could not provide any detailed taxonomic information but reported that radiolarian 'colonies' were 45 cm wide and 12 cm high. In their opinion, radiolarians contributed substantially to the planktonic biomass in the area.

As to the radiolarian remains in the surface sediment, the seafloor in the Polish contract area does not generally feature siliceous ooze, but Nigrini (1967) reported on radiolarian found in the sediment cores from a few areas in the vicinity (Table 3).

Table 2 Radiolarian taxa present in the water column at 20°55.3′N; 19°44.5′W (Boltovskoy et al. 1996) which may occur in the Polish contract area on the MAR

Radiolarian taxa	
Arachnocorallium caivata a	*Hexacontium armatum/hostiie*
Arachnocorallium caivata e	*Lamprocyclas hannai*
Tetraplecta pinigera	*Actinomma leptodermum*
Botryocephalina armata	*Hexalonche aristarchi*
Siphocampe arachnea	*Triceraspyris antarctica*
Artostrobus joergenseni	*Botryostrobus auritus/australis*
Artostrobus annulatus	*Trisulcus triacanthus*
Heiotholus histricosa	*Lophophaena buetschlii*
Actinomma sp. 1	*Pseudocubus obeliscus*

Table 3 Abundance of radiolarian remains in the Atlantic Ocean sediment sampled in the vicinity of the Polish contract area (Nigrini 1967)

Coordinates	Water depth (m)	Abundance and preservation state of radiolarian remains
29°35′N, 42°33′W	2820	Moderately abundant, well-preserved
21°05′N, 58°50′W	5596	Very rare
34°54′N, 44°20′W	4281	Abundant, well-preserved
33°07′N, 52°14′W	5651	Very rare

2 The Mid-Atlantic Ridge Zooplankton

The zooplankton described here comprise animals ranging in size from 0.2 up to 20 mm. Generally, the zooplankton can be subdivided into two major groups: the holoplankton (e.g., copepods and krill) which spend their entire lives as plankton drifting with ocean currents, and thus provide major food sources for higher trophic levels and act as major contributors to the oceanic 'biological pump'; and the meroplankton (e.g., larvae of barnacles, mussels, annelids and fish) which spend only part of their lives in the pelagial.

The zooplankton are difficult to sample adequately because individual organisms are tiny and often patchily distributed. Moreover, some zooplankton species that live mainly in the water between 500 and 1000 m deep perform massive diel vertical migrations (DVM), ascending the surface layers of the ocean at night, where they feed on plankton, and descending at dawn, to digest the food and to minimise the predation risk which is at its highest in the upper water layers. Until recently, it had been thought that DVMs were triggered by a relative change in the visible light intensity. However, acoustic data, a measure of zooplankton abundance and movement in the North-Atlantic Ocean in the vertical range of 500–1600 m, provided evidence that DVMs also occur in the deep sea where no direct and background sunlight penetrates (van Haren and Compton 2013). Acoustic Doppler current profilers were moored at a location similar to the MAR at eight different sub-tropical positions (from 33°N to the Equator and from 22°W to 37°W), and the migrations precisely followed the rhythm of local sunrise and sunset between 500 and 650 m. They continued below 650 m where the deepest penetrable irradiance levels are $<10^{-7}$ times their near-surface values, but the plankton shortened their time at depth by up to about 63% at 1600 m. The trend was consistent both across the latitudes investigated and between different seasons. It was thus hypothesized that the solar diurnal and seasonal rhythms in the deep sea plankton migrations could be controlled by a mechanism other than light, e.g., a precise biochemical clock. In accordance with this hypothesis, the deepest plankton were consistently the first to migrate upwards (van Haren and Compton 2013).

Deep-sea zooplankton communities are generally very diverse, but occur at low abundances. Thus, large sampling systems are needed to filter sufficient amounts of water (Vinogradov et al. 1996). In consequence, the zooplankton of the subtropical-tropical regions are generally very poorly studied. The first comprehensive assessment of the zooplankton community in the Atlantic Ocean at low latitudes (35°N–40°S, near the Polish contract area) covered a broad depth range (0–3000 m) (de Puelles et al. 2019). The study demonstrated that the zooplankton abundance declined sharply with depth, comprising, on average, 82% of the depth-integrated abundance in the epipelagic layer. The mesopelagic zooplankton contributed 4 to 12% of the water column abundance, while the bathypelagic zooplankton accounted for <1% of the total abundance. Tropical-subtropical zooplankton abundances reported by de Puelles et al. (2019) were generally low, usually amounting to >200 ind./m^3 in the epipelagic layer, around 100 ind./m^3 in the mesopelagic layer, and < 3 ind./m^3 in the bathypelagic zone. These low values might be an artefact of the relatively large mesh size (>300 μm) of the plankton net used, and thus an under-representation of the numerically important tiny copepods. However, similar values were found in the upper 300 m when a smaller (150 μm) mesh size net was used along a latitudinal transect in the eastern Atlantic from 34°49.5′N to 27°28.1′S, but further east (Schnack-Schiel et al. 2010) than the stations sampled by de Puelles et al. (2019).

Zooplankton communities in subtropical-tropical regions are numerically dominated by small-sized copepods, particularly calanoids of the families Clausocalanidae and Paracalanidae and cyclopoids, often contributing jointly more than 50% to the total number of copepods (e.g., Schnack-Schiel et al. 2010). They are generally known to be well adapted to low phytoplankton concentrations in oligotrophic waters. De Puelles et al. (2019) identified seventeen different, dominant zooplankton groups, of which seven displayed contributions <1%. Overall, copepods were the dominant group in all the samples (80%), followed by chaetognaths (5%), ostracods (3%), and siphonophores (3%). Other groups such as appendicularians (2%), euphausiids (1%), and amphipods (1%) were rarely observed (de Puelles et al. 2019). The high dominance of copepods confirms their key role in the marine pelagic food web—in the transfer of primary production and microzooplankton biomass to higher trophic levels (Vinogradov et al. 1996). The study by de Puelles et al. (2019) clearly demonstrated that smaller copepods were usually abundant in the upper strata, while large copepods were mainly distributed in deeper waters.

Schnack-Schiel et al. (2010) identified a total of 149 calanoid copepod species in the upper 300 m of their area of study. They pointed to *Clausocalanus* as the most abundant genus, accounting for an average of about 45% of all calanoids. It was followed by *Calocalanus* (13%), *Delibus* (9%), *Paracalanus* (6%), and *Pleuromamma* (5%). The list of the most abundant cosmopolitan cyclopoid genera in the area was topped by *Oithona*, followed by *Oncaea*, both widely distributed throughout the low latitudes, particularly in the upper 300 m.

The study of de Puelles et al. (2019), carried out in the epipelagic zone of subtropical-tropical Atlantic [35°N–40°S but further west from the area sampled by Schnack-Schiel et al. (2010)] showed, too, the zooplankton to be numerically dominated by small-sized cosmopolitan species such as *Clausocalanus furcatus*, *Nannocalanus minor* and *Oithona plumifera*. The upper mesopelagial (200–500 m) was inhabited mainly by three representatives of *Pleuromamma* (*P. abdominalis*, *P. gracilis* and *P. piseki*), while the lower part of the zone (500–1000 m) featured *Metridia brevicauda*, *Conaea* sp. and *Pleuromamma xiphias*.

Dominants similar to those in the lower mesopelagial were characteristic of the bathypelagic zone (>1000 m depth), with a substantial contribution of *Triconia conifera* (de Puelles et al. 2019). The mesopelagic zone dominants (*P. abdominalis*, *P. gracilis*, *P. piseki*, and *P. xiphias*), reported as the daytime inhabitants of the deeper layer, are known to be the strong vertical migrants (e.g., Gaard et al. 2008).

The high species diversity found by de Puelles et al. (2019) and Schnack-Schiel et al. (2010) is a common feature of the tropical and subtropical domains. The diversity peaks in mid-waters, and the species richness declines with depth, which demonstrates the global nature of such patterns, earlier reported in regional assessments of planktonic ostracods (e.g., Angel et al. 2007). The tropical zooplankton communities are characterized by a high species richness and small changes of biomass throughout the year (e.g., Longhurst 1993; Bode et al. 2018; Schnack-Schiel et al. 2010; Vereshchaka et al. 2017).

The horizontal distribution of planktonic organisms in tropical latitudes is more or less unrestricted due to the absence of physical barriers, thus allowing the co-occurrence and wide latitudinal ranges of many oceanic species (e.g., Schnack-Schiel et al. 2010). In contrast, a strong vertical stratification, i.e., a pronounced thermocline and oxycline, effectively separates the mixed, well oxygenated, warm surface layer from cooler, less oxygenated waters below. Hence, zooplankton communities may show differences in their vertical distribution, consistent with species-specific ecophysiological preferences; however, depth and temperature are usually the most important factors structuring their populations throughout the water column (de Puelles et al. 2019). A strong vertical zonation of the zooplankton was reported by Vereshchaka and Vinogradov (1999) from visual censuses at the Broken Spur hydrothermal field (in the Polish contract zone on the MAR). They found the plankton biomass to form 2 major aggregations, one within the main pycnocline (~1000 m depth) and the other near the hydrothermal fluid plume, both aggregations being dominated by the gelatinous zooplankton (medusae, ctenophores, siphonophores) and radiolarians, in addition to the abundant copepods.

The Lost City, another hydrothermal field in the Polish contract area on the MAR, was surveyed for the zooplankton (visual census from a submersible) by Vinogradov et al. (2003). In their study, a characteristic and common feature of the site, not encountered elsewhere, was the accumulations of the macroplankton and nektobenthos.

The study of zooplankton on hydrothermal vent system-supporting mid-oceanic ridges such as the MAR carries a benefit of contributing to the assessment of connectivity between different areas, particularly important for conservation and management purposes. The connectivity is provided primarily by the transport of benthic organisms' larvae, i.e., the meroplankton (e.g., Comtet et al. 2000; Beaulieu et al. 2009). In this regard, mention has to be made of the *Rimicaris* sp. shrimp larvae observed by Vinogradov et al. (2003) over the Broken Spur hydrothermal field in the layers closest to the seafloor.

3 Macroplankton and Micronekton

These categories of pelagic organisms group water column inhabitants larger than 5 mm in size, both passive floaters (the majority of gelatinous plankton) and active swimmers (cephalopods, shrimps, krill and amphipods). They occur throughout the water column, from the surface to the near-bottom water layer (benthopelagic crustaceans). Due to their size, the organisms dealt with here are of special importance for large predators, from fish to seabirds and sea mammals. A number of macroplankton and micronekton species tend to occur in aggregations; they form swarms and shoals, due to both physical forcing (aggregations on the density interface, internal waves or eddies) and biological (behavioural) reactions (mainly krill, shrimps, and cephalopods).

The pelagic area of concern (central North Atlantic) belongs to the subtropical zone, one of the biodiversity-richest among open ocean regions (McRae et al. 2017), and is inhabited by a mixture of boreal and tropical faunal elements (Judkins and Haedrich 2018). The dominant non-fish pelagic organisms include 65 shrimp species (Cardoso et al. 2014), followed by 41 species of cephalopods (Vecchione et al. 2010a, b). The benthopelagic carrion-feeding amphipods are represented by over 35 species (Horton et al. 2013); other pelagic Amphipoda are the Hyperiida represented by several mesopelagic species (Bellan-Santini and Thurston 1996; Burukovsky and Falkenhaug 2015), with a generally very wide distribution in the deep Atlantic.

The jellyfish, representing the gelatinous plankton and important for their role in the food web, are widespread over the central Atlantic, and count at least 51 species (scyphozoans, cnidarians, and appendicularians), most feeding on the microplankton and mesozooplankton (Stemmann et al. 2008; Hosia et al. 2017).

As regards the macroplankton and non-fish micronekton at individual hydrothermal vent fields, from the standpoint of the Polish exploration at the MAR, important are data of Russian researchers, e.g., Vinogradov et al. (2003) who conducted visual censuses of organisms in the water column from a manned submersible (known also as a human-operated vehicle, HOV) MIR 1 at the Lost City vent field in the Polish contract area. In their study, a characteristic and common feature of the site, not encountered elsewhere, was the accumulation of the euphausiids *Nematoscelis* aff. *tenellae* (preying on copepods) in the near-bottom layer as well as the presence of free-swimming amphipods *Paraphronima crassipes*, *Streetsia challengeri*, and *Primno* sp. In addition, there were abundant benthic amphipods *Bouvierella* aff. *curtirama* popping up into the water column.

Further data on the large zooplankton of both the Broken Spur and Lost City (supplied by visual censuses and net hauls) were discussed by Vinogradov and Vereshchaka (2006).

Table 4 shows the main macroplanktonic species recorded in the central part of MAR; usually, the most common species are also most important for the food web structure and parasite transmission to the top predators. The conservation value of the listed species is difficult to assess, as this whole ecological group is data-deficient and potentially widespread both over the Atlantic and the depth zones as well.

Table 4 Principal non-fish micronekton and macroplankton species in the area (compiled from the literature cited)

Species	Higher taxon	Depth strata (m)	Remarks
Aeginura grimaldii	Scyphozoa	500–800	Most common gelatinous zooplankter in the area
Lensia subtilis	Siphonophora	Pelagic, 100–1000 m	Common cosmopolitan species
Abyssorchomene abyssorum	Amphipoda	Near-bottom waters	Most common necrophagic amphipod in the area
Paraphronima crassipes	Amphipoda	Near-bottom waters	Lost City (Polish contract area)
Streetsia challengeri	Amphipoda	Near- bottom waters	Lost City (Polish contract area)
Primno sp.	Amphipoda	Near-bottom waters	Lost City (Polish contract area)
Phronima sedentaria	Hyperiida	Mesopelagic (500–1000)	Cosmopolitan common species
Meganyctiphanes norvegica	Euphausiacea	Pelagic (100–1000)	Common, northern element
Nematoscelis aff. *tenellae*	Euphausiacea	Near bottom waters	Lost City (Polish contract area)
Acanthephyra pelagica	Decapoda	Mesopelagic, bathypelagic (500–3500)	Most common pelagic shrimp in the area
Gonatus steenstrupi	Cephalopoda	100–1000	Most common squid in the northern sector of the area

4 Fishes

As there is no published information about any demersal fishing activities in the Polish contract area of the Mid-Atlantic Ridge (MAR), there are pretty good chances that bottom fishing is not commercially feasible there. Contrary to that, the epipelagic waters of the MAR area in question were under a very moderate pressure from the long-line tuna, billfish and shark fisheries, but free of any purse seine or bait boat fishing activities, according to data on the fishing effort allocated to five-degree statistical squares and published up to 2010 (ICCAT Stat. Bull. https://www.iccat.int/en/scrs.html).

The Polish contract area is located approximately within the two ICCAT statistical squares, i.e., 25°–35°N and 40°–45°W. On the basis of the reported commercial catches per those squares, the following large pelagic fish species are expected to be the most frequent visitors in the area: albacore (*Thunnus alalunga*), bigeye (*T. obesus*), northern bluefin tuna (*T. t. thynnus*), blue marlin (*Macaira nigricans*), sailfish (*Istiophorus albicans*), white marlin (*Terapturus albidus*), swordfish (*Xiphias gladius*), blue shark (*Prionace glauca*), and shortfin mako (*Isurus oxyrinchus*). They are apex predators which feed on a variety of smaller fish, cephalopods, and crustaceans which are distributed in a wide depth range of epi- and mesopelagic zones. Therefore, some of them, e.g., billfish and swordfish, but also certain tuna species, perform diurnal feeding migrations while, e.g,. sailfish stay at shallower depths (Carey and Robison 1981; Dagorn et al. 2000; Lerner et al. 2013; Bubley et al. 2020).

Food resources of the apex predators do not consist of mesopelagic species only, but include also epipelagic ones, e.g., flying fish and Atlantic saury (*Scomberesox saurus*) (Biton Porsmoguer et al. 2014). Although the major prey composition of the apex species mentioned differs from region to region, in the open oceanic waters the mesopelagic organisms constitute their important prey (Moteki et al. 2001; Satoh et al. 2004). Therefore, the vertically migrating mesopelagic fish, mainly numerous species of the family Myctophidae, but also species of the family Gonostomatidae which perform vertical migrations, play an important role in the vertical energy transfer process in the oceanic ecosystems. The most abundant *Cyclotone* spp. (Gonostomatidae) are too small as a prey for apex predators and do not perform vertical migrations (Badcock and Merrett 1976), but certainly are utilized as prey by much smaller mesopelagic predators.

Thus, informationon on mesopelagic fishes within the study area is important. As the relevant data for the Polish contract area on the MAR are missing, the most representative information on the mesopelagic fish abundance, species composition, and vertical distribution could be extracted from datasets collected within the same mesopelagic ecoregion, i.e., the Northeastern Central Atlantic sub-ecoregion (Judkins and Haedrich 2018), or from the geographically closest sampling sites. The primary factor determining the pelagic fish assemblage composition in the northern part of the MAR (from Iceland to the Azores) is the depth, while the geographic region is of secondary importance (Sutton et al. 2008).

In addition to the results from the southernmost part of the area studied by Sutton et al. (2008), the mesopelagic fish structure close to the border between the Mauretania-Cape Verde ecoregion and Northeastern Central Atlantic (Badcock and Merrett 1976) and Northern Sargasso Sea and Southern Sargasso Sea subregions (Sutton et al. 2010), could provide proxy information on the mesopelagic fishes assemblage in the Polish contract area on the MAR. For example, the 33 stratified samples collected with RMT 1+8 net between the surface and 2010 m depth showed the following peaks of mesopelagic fish abundance: one peak in the 405–600 m depth stratum during the day and two peaks within 100–200 m and 405–500 m during the night (Badcock and Merrett 1976). The species composition was highly dominated by small *Cyclothone* species (Gonostomatidae), i.e., *C. braueri* (9–38 mm, occurring in the 400–600 m depth stratum during day and night), and *C. microdon* (11–59 mm, occurring in the 600–1250 m depth stratum during the day and 500–2000 m during the night). *Vallencienellus tripunctulatus* and *Argyropelecus hemigymnus* (Sternoptychidae) were the third and fourth most numerous species, respectively, while the two most abundant myctophids (*Benthosema suborbitale* and *Hygophum hygomii*) were far less numerous than the gonostomatid and sternoptychid species (Badcock and Merrett 1976) mentioned above. The samples analyzed by Badcock and Merrett (1976) included 37 taxa of stomiatoid fishes (suborder Stomiatoidei), while the single family Myctophidae was represented by 31 species.

In the Azorean Zone (0–750 m), 29 myctophid species contributed half (50.8%) of the total assemblage number and one-quarter of the biomass (26.9%) (Sutton et al. 2008) with the myctophid *Lobianchia dofleini* dominating (18.6%), and the gonostomatids *C. microdon* (14.2%) and *C. braueri* (8.1%) taking the second and third place, respectively, in the overall number of the specimens collected, respectively. Samples collected in the Sargasso Sea (0–1000 m depth stratum) differed from those reported in papers mentioned earlier in that four *Cyclotone* species were the most abundant ones (*C. braueri*, *C. pallida*, *C. microdo*n and *C. pseudopalida* with 47.5, 19.9, 5.9 and 3.2%, respectively), while the first myctophid species was the fifth (*Lepidophanes guentheri* with 2.5%) (Sutton et al. 2010).

Cyclotone spp. and myctophid species were reported by Vereshchaka and Vinogradov (1999) from the area of the Broken Spur vent field in the Polish contract area on the MAR. The authors found the fish mentioned, with a biomass of 3 to 40 mg/m^3, to inhabit primarily the main pycnocline (~1000 m depth). *Cyclotone* spp. were also observed in the vicinity of the hydrothermal fluid discharge. Myctophids were present, too, within the main pycnocline, with vertical ranges overlapping those of *Cyclotone* spp.

5 Marine Birds, Mammals and Turtles

This section provides a list of species and basic information about the distribution, abundance, breeding and conservation status of pelagic seabirds, cetaceans and sea turtles observed in the Mid-Atlantic Ridge area; the list does not contain sea snakes, as these are not found here.

The attached tables (with information extracted from the references listed at the end of the section) cover 60 species of seabirds (including 19 breeding [Br]), 28 cetaceans and 5 sea turtles (including 4 breeding), for which the closest breeding sites are Cape Verde in the north and Saint Helena and Ascension islands in the south (Hazevoet 1995; Ashmole and Ashmole 2000; McCulloch 2004; Prater 2012). The others live in the area all year round—they are non-breeding species or their breeding has not been confirmed here (Harrison 1983). This group also includes species that winter here and migrate through the area seasonally. However, in the tropics, with the exception of a few migratory species, seasonal changes in the abundance of birds, mammals and sea turtles are hardly noticeable.

The Mid-Atlantic Ridge area hosts several species of greatest concern, including two endemic (E) species (the Ascension frigate bird and the Atlantic humpback dolphin), two critically endangered (CR) species (the Atlantic humpback dolphin and the hawksbill sea turtle), as well as four vulnerable (VU) seabirds, one cetacean and three species of sea turtles (Klinowska and Cooke 1991; Birdlife International 2000, 2012; Fishpool and Evans 2001; Wilson and Reeder 2005) (see Tables 5, 6, and 7).

Table 5 Seabirds found in the tropical/Mid-Atlantic region

Taxonomic position	English name	Latin name	Conservation status	Comments
Diomedeidae	Wandering albatross	*Diomedea exulans*	VU	A
	Black-browed albatross	*Thalassarche melanophris*	LC	A
	Yellow-nosed albatross	*Thalassarche chlororhynchos*	EN	A
	Sooty albatross	*Phoebetria fusca*	EN	A
Procellariidae	Antarctic giant petrel	*Macronectes giganteus*	LC	A
	Southern fulmar	*Fulmarus glacialoides*	LC	A
	Cape petrel	*Daption capense*	LC	A
	Fea's petrel	*Pterodroma feae*	NT	(Br) Cape Verde & Madeira
	Murphy's petrel	*Pterodroma ultima*	NT	A
	Soft-plumaged petrel	*Pterodroma mollis*	LC	A
	Broad-billed prion	*Pachyptila vittata*	LC	A
	Bulwer's petrel	*Bulweria bulwerii*	LC	(Br) Cape Verde & St Helena
	White-chinned petrel	*Procellaria aequinoctialis*	VU	A

(continued)

Table 5 (continued)

Taxonomic position	English name	Latin name	Conservation status	Comments
	Cory's shearwater	*Calonectris borealis*	LC	A
	Scopoli's shearwater	*Calonectris diomedea*	LC	A
	Cape Verde shearwater	*Calonectris edwardsii*	NT	(Br) Cape Verde & Canary I.
	Great shearwater	*Ardenna gravis*	LC	
	Sooty shearwater	*Ardenna griseus*	NT	A
	Manx shearwater	*Puffinus puffinus*	LC	A
	Boyd's shearwater	*Puffinus boydi*	EN	(Br) Cape Verde
	Barolo shearwater	*Puffinus baroli*	Ex	A; (Br) north of Canary Islands
Oceanitidae	Wilson's storm petrel	*Oceanites oceanicus*	LC	
	White-faced storm petrel	*Pelagodroma marina*	LC	(Br) Cape Verde
	White-bellied storm petrel	*Fregetta grallaria*	LC	A
	Black-bellied storm petrel	*Fregetta tropica*	LC	A
Hydrobatidae	European storm petrel	*Hydrobates pelagicus*	LC	
	Cape Verde storm petrel	*Oceanodroma jabejabe*	EN	(Br) Cape Verde
	Band-rumped storm petrel	*Oceanodroma Castro*	LC	(Br) Cape Verde & Ascension I.
	Leach's storm petrel	*Oceanodroma leucorhoa*	LC	
Phaethontidae	Red-billed tropicbird	*Phaethon aethereus*	LC	(Br) Cape Verde, St Helena & Ascension I.
	White-tailed tropicbird	*Phaethon lepturus*	LC	(Br) Ascension I.
Sulidae	Northern gannet	*Morus bassanus*	LC	A
	Masked booby	*Sula dactylatra*	LC	(Br) Ascension I.
	Red-footed booby	*Sula sula*	LC	(Br) Ascension I.
	Brown booby	*Sula leucogaster*	LC	(Br) Cape Verde, St Helena & Ascension I.

(continued)

Table 5 (continued)

Taxonomic position	English name	Latin name	Conservation status	Comments
Fregatidae	Magnificent frigatebird	*Fregata magnificens*	LC	(Br) Cape Verde
	Great frigatebird	*Fregata minor*	LC	
	Lesser frigatebird	*Fregata ariel*	LC	
	Ascension frigatebird	*Fregata aquila*	VU	(E), (Br) Ascension I.
Stercorariidae	Great skua	*Catharacta skua*	LC	
	Pomarine jaeger	*Stercorarius pomarinus*	LC	
	Parasitic jaeger	*Stercorarius parasiticus*	LC	
	Long-tailed jaeger	*Stercorarius longicaudus*	LC	A
	Brown skua	*Stercorarius antarctica*	LC	A
Laridae	Lesser black-backed gull	*Larus fuscus*	LC	
	Caspian gull	*Larus cachinnans*	LC	
	Yellow-legged gull	*Larus michahellis*	LC	
	Black-headed gull	*Chroicocephalus ridibundus*	LC	
	Slender-billed gull	*Chroicocephalus genei*	LC	A
	Sabine's gull	*Xema sabini*	LC	
	Black-legged kittiwake	*Rissa tridactyla*	VU	
	Roseate tern	*Sterna dougallii*	LC	
	Common tern	*Sterna hirundo*	LC	
	Arctic tern	*Sterna paradisaea*	LC	
	Sooty tern	*Onychoprion fuscatus*	LC	(Br) Ascension I.
	Brown noddy	*Anous stolidus*	LC	(Br) Ascension I.
	Black noddy	*Anous minutus*	LC	(Br) Ascension I.
	White tern	*Gygis alba*	LC	(Br) St Helena & Ascension I.
	Antarctic tern	*Sterna vittata*	LC	A
Alcidae	Common guillemot	*Uria aalge*	LC	

Table 6 Cetaceans reported from the tropical/Mid-Atlantic region

Taxonomic position	English name	Latin name	Conservation status	Comments
Mysticeti				
	Fin whale	*Balaenoptera physalus*	VU	
	Humpback whale	*Megaptera novaeangliae*	LC	
	Bryde's whale	*Balaenoptera brydei*	LC	
	Sei whale	*Balaenoptera borealis*	EN	
	Blue whale	*Balaenoptera musculus*	EN	
	Northern minke whale	*Balaenoptera acutorostrata*	LC	
Odontoceti				
Delphinidae	Short-beaked common dolphin	*Delphinus delphis*	LC	
	Long-beaked common dolphin	*Delphinus capensis*	DD	Shelf areas
	Fraser's dolphin	*Lagenodelphis hosei*	LC	
	Atlantic humpback dolphin	*Sousa teuszi*	CR	(E) W Africa shelf
	Pantropical spotted dolphin	*Stenella attenuate*	LC	
	Atlantic spotted dolphin	*Stenella frontalis*	LC	
	Spinner dolphin	*Stenella longirostris*	LC	
	Clymene dolphin	*Stenella clymene*	LC	
	Striped dolphin	*Stenella coeruleoalba*	LC	
	Common bottlenose dolphin	*Tursiops truncates*	LC	
Orcininae	Pygmy killer whale	*Feresa attenuate*	LC	
	Short-finned pilot whale	*Globicephala macrorhynchus*	LC	
	Killer whale (orca)	*Orcinus orca*	DD	
	Melon-headed whale	*Peponocephala electra*	LC	
	False killer whale	*Pseudorca crassidens*	NT	
	Rough-toothed dolphin	*Steno bredanensis*	LC	
Physeteroidea	Dwarf sperm whale	*Kogia sima*	LC	
	Pygmy sperm whale	*Kogia breviceps*	LC	
	Sperm whale	*Physeter macrocephalus*	EN	Dives to 2250 m

(continued)

Table 6 (continued)

Taxonomic position	English name	Latin name	Conservation status	Comments
Ziphioidea	Gervais' beaked whale	*Mesoplodon europaeus*	DD	
	Blainville's beaked whale	*Mesoplodon densirostris*	DD	
	Cuvier's beaked whale	*Ziphius cavirostris*	LC	

Table 7 Sea turtles known from the tropical/Mid-Atlantic region

Taxonomic position	English name	Latin name	Conservation status	Comments
Carettinae	Loggerhead sea turtle	*Caretta caretta*	VU	Br
	Olive ridley sea turtle	*Lepidochelys olivacea*	VU	
Cheloniinae	Green sea turtle	*Chelonia mydas*	EN	Br
	Hawksbill sea turtle	*Eretmochelys imbricata*	CR	Br
Dermochelyinae	Leatherback	*Dermochelys coriacea*	VU	Br

IUCN Red List of Threatened Species: *CR*, critically endangered; *EN*, endangered; *VU*, vulnerable; *LC*, least concern; *(Br)*, breeding; *(A)*, accidental—occurs rarely or accidentally; *(E)*, endemic

References

Afonso P, McGinty N, Graca G, Fontes J, Inacio M, Totland A, Menezes G (2014) Vertical migrations of a deep-sea fish and its prey. PLoS One 9:e97884. https://doi.org/10.1371/journal.pone.0097884

Agusti S, González-Gordillo J, Vaqué D (2015) Ubiquitous healthy diatoms in the deep sea confirm deep carbon injection by the biological pump. Nat Commun 6:7608. https://doi.org/10.1038/ncomms8608

Allen J, Brown L, Sanders R (2005) Diatom carbon export enhanced by silicate upwelling in the Northeast Atlantic. Nature 437:728–732. https://doi.org/10.1038/nature03948

Anderson RO (1983) Radiolaria. Springer, New York

Anderson TR, Martin AP, Lampitt RS, Trueman CN, Henson SA, Mayor DJ (2019) Quantifying carbon fluxes from primary production to mesopelagic fish using a simple food web model. ICES J Mar Sci 76:690–701. https://doi.org/10.1093/icesjms/fsx234

Angel MV, Błachowiak-Samołyk K, Drapun I, Castillo R (2007) Changes in the composition of planktonic ostracod populations across a range of latitudes in the North-East Atlantic. Prog Oceanogr 73:60–78

Ashmole NP, Ashmole MJ (2000) St Helena and Ascension Island: a natural history. Anthony Nelson, Oswestry, UK

Badcock J, Merrett NR (1976) Midwater fishes in the eastern North Atlantic-I. Vertical distribution and associated biology in 30°N, 23°W, with developmental notes on certain myctophids. Prog Oceanogr 7:3–58. https://doi.org/10.1016/0079-6611(76)90003-3

Barton AD, Lozier MS, Williams RG (2015) Physical controls of variability in North Atlantic phytoplankton communities. Limnol Oceanogr 60:181–197. https://doi.org/10.1002/lno.10011

Beaulieu SE, Mullineaux LS, Adams DK, Mills SW (2009) Comparison of a sediment trap and plankton pump for timeseries sampling of larvae near deep-sea hydrothermal vents. Limnol Oceanogr Methods 7:235–248

Bellan-Santini D, Thurston MH (1996) Amphipoda of the hydrothermal vents along the Mid-Atlantic Ridge. J Nat Hist 30:685–702. https://doi.org/10.1080/00222939600770381

Birdlife International (2000) Threatened birds of the world. Lynx Edicions and BirdLife International, Barcelona and Cambridge, UK

BirdLife International (2012) *Calonectris edwardsii*. IUCN Red List of Threatened Species

Biton Porsmoguer S, Banaru D, Bearez P, Dekeyser I, Merchan Fornelino M, Boudouresque CF (2014) Unexpected headless and tailless fish in the stomach content of shortfin Mako *Isurus oxyrinchus*. PLoS One 9:e88488. https://doi.org/10.1371/journal.pone.0088488

Bjørklund KR, Cortese G, Swanberg N, Schrader HJ (1998) Radiolarian faunal provinces in surface sediments of the Greenland, Iceland and Norwegian (GIN) seas. Mar Micropaleontol 35:105–140

Bode M, Hagen W, Cornils A, Kaiser P, Auel H (2018) Copepod distribution and biodiversity patterns from the surface to the deep sea along a latitudinal transect in the eastern Atlantic Ocean (24°N to 21°S). Prog Oceanogr 161:66–77

Boltovskoy D, Uliana E, Wefer G (1996) Seasonal variation in the flux of microplankton and radiolarian assemblage compositions in the northeastern tropical Atlantic at 2,195 m. Limnol Oceanogr 41(4):615–635

Bourne WRP (2008) The birds of the Cape Verde Islands. Ibis 97:508–556. https://doi.org/10.1111/j.1474-919X.1955.tb04981.x

Bubley WJ, Galuardi B, Dukes AW, Jenkins WE (2020) Incorporating depth into habitat use descriptions for sailfish *Istiophorus platypterus* and habitat overlap with other billfishes in the western North Atlantic. Mar Ecol Prog Ser 638:137–148

Burukovsky RN, Falkenhaug T (2015) Feeding of the pelagic shrimp *Acanthephyra pelagica* (Risso, 1816) (Crustacea: Decapoda: Oplophoridae) in the northern Mid-Atlantic Ridge area in 1984 and 2004. Arthropoda Sel 24:303–316

Cardoso IA, Falkenhaug T, Fernandes LFL (2014) Comparison between the shrimp species richness (Caridea and Dendrobranchiata, Decapoda, Crustacea) of the south and north Mid-Atlantic Ridge. Braz J Oceanogr 62:47–56

Carey FG, Robison BH (1981) Daily patterns in the activities of sword-fish, *Xiphias gladius*, observed by acoustic telemetry. U.S. National Marine Fisheries Service Fishery Bulletin, vol 79, pp 277–292

Comtet T, Jolliver D, Khripounoff A, Segonzac M, Dixon D (2000) Molecular and morphological identification of settlement-stage vent mussel larvae, *Bathymodiolus azoricus* (Bivalvia: Mytilidae), preserved in situ at active vent fields on the Mid-Atlantic Ridge. Limnol Oceanogr 45:1655–1661

Dagorn L, Bach P, Josse E (2000) Movement patterns of large bigeye tuna (*Thunnus obesus*) in the open ocean, determined using ultrasonic telemetry. Mar Biol 136:361–371. https://doi.org/10.1007/s002270050694

de Master DJ (1981) The supply and accumulation of silica in the marine environment. Geochim Cosmochim Acta 45:1715–1732

de Puelles MLF, Gazá M, Cabanellas-Reboredo MS, Santandreu MM, Irigoien X, González-Gordillo JI, Duarte CM, Hernández-León S (2019) Zooplankton abundance and diversity in the tropical and subtropical ocean. Diversity 11:203

De Wever P, Dumitrica P, Caulet JP, Nigrini C, Caridroit M (2001) Radiolarians in the sedimentary record. Gordon & Breach Science Publishers

Fishpool LDC, Evans MI (2001) Important bird areas in Africa and Associated Islands: priority sites for conservation. Pisces Publications and BirdLife International (BirdLife Conservation Series no. 11). Newbury and Cambridge, UK

Gaard E, Gislason A, Falkenhaug T, Søiland H, Musaeva E, Vereshchaka A, Vinogradov G (2008) Horizontal and vertical copepod distribution and abundance on the Mid-Atlantic Ridge in June 2004. Deep Sea Res Part II Top Stud Oceanogr 55:59–71

Haq BU, Boersma A (1998) Introduction to marine micropaleontology. Elsevier

Harrison P (1983) Seabirds - an identification guide. Croom Helm, London

Hazevoet CJ (1995) The birds of the Cape Verde Islands. British Ornithologists' Union (Check-list 13), Tring, UK

Horton T, Thurston MH, Duffy GA (2013) Community composition of scavenging amphipods at bathyal depths on the Mid-Atlantic Ridge. Deep Sea Res Part II Top Stud Oceanogr 98:352–359

Hosia A, Falkenhaug T, Baxter EJ, Pagès F (2017) Abundance, distribution and diversity of gelatinous predators along the northern Mid-Atlantic Ridge: a comparison of different sampling methodologies. PLoS One 12:0187491. https://doi.org/10.1371/journal.pone.0187491

IUCN (2015) The IUCN Red List of Threatened Species. IUCN

Judkins D, Haedrich R (2018) The deep scattering layer micronektonic fish faunas of the Atlantic mesopelagic ecoregions with comparison of the corresponding decapod shrimp faunas. Deep Sea Res Part I Oceanogr Res Pap 136:1–30. https://doi.org/10.1016/j.dsr.2018.04.008

Klinowska M, Cooke J (1991) Dolphins, porpoises, and whales of the world: the IUCN Red Data Book. IUCN Publications, New York

Kotrys B (2012) Palaeotemperature reconstructions at the polar front in the Southern Ocean (Atlantic sector) based on middle–late Pleistocene radiolarian record from ODP Leg 177 Site 1091. Ph.D. Thesis, University of Szczecin

Lerner JD, Kerstetter DW, Prince ED, Talaue-McManus L, Orbesen ES, Mariano A, Snodgrass D, Thomas GL (2013) Swordfish vertical distribution and habitat use in relation to diel and lunar cycles in the Western North Atlantic. Trans Am Fish Soc 142:95–104

Lindsay D, Umetsu M, Grossmann M, Miyake H, Yamamoto H (2015) The gelatinous macroplankton community at the Hatoma Knoll Hydrothermal Vent. In: Ishibashi J, Okino K, Sunamura M (eds) Subseafloor biosphere linked to hydrothermal systems. Springer, Tokyo. https://doi.org/10.1007/978-4-431-54865-

Longhurst A (1993) Seasonal cooling and blooming in tropical oceans. Deep Sea Res Part I Oceanogr Res Pap 40:2145–2165

Maynard NG (1976) Relationship between diatoms in surface sediments of the Atlantic Ocean and the biological and physical oceanography of overlying waters. Paleobiology 2:99–121

McCulloch N (2004) A guide to the birds of St Helena and Ascension Island. Royal Society for the Protection of Birds

McRae L, Deinet S, Freeman R (2017) The diversity-weighted living planet index: controlling for taxonomic bias in a global biodiversity indicator. PLoS One 12:e0169156. https://doi.org/10.1371/journal.pone.0169156

Moteki M, Arai M, Tsuchiya K, Okamoto H (2001) Composition of piscine prey in the diet of large pelagic fish in the eastern tropical Pacific Ocean. Fish Sci 67:1063–1074

Nigrini C (1967) Radiolaria in pelagic sediments from the Indian and Atlantic Oceans. Bull Scripps Inst Oceanogr 11:1–125

Norton TA, Melkonian M, Andersen RA (1996) Algal biodiversity. Phycologia 35:308–326

Oksman M, Juggins S, Miettinen A, Witkowski A, Weckström K (2019) The biogeography and ecology of common diatom species in the northern North Atlantic, and their implications for paleoceanographic reconstructions. Mar Micropaleontol 148:1–28. https://doi.org/10.1016/j.marmicro.2019.02.002

Olguin HF, Boltovskoy D, Lange CB, Brandini F, Flyn KJ (2006) Distribution of spring phytoplankton (mainly diatoms) in the upper 50 m of the southwestern Atlantic Ocean (30–61°S). J Plankton Res 28:1107–1128. https://doi.org/10.1093/plankt/fbl045

Prater T (2012) Important bird area: St Helena. Br Birds 105:638–653

Satoh K, Yokawa K, Saito H, Matsunaga H, Okamoto H, Uozumi Y (2004) Preliminary stomach contents analysis of pelagic fish collected by Shoyo-Maru 2002 research cruise in the Atlantic Ocean. Collect Vol Sci Pap ICCAT 56:1096–1114

Schnack-Schiel SB, Mizdalski E, Cornils A (2010) Copepod abundance and species composition in the eastern subtropical/tropical Atlantic. Deep Sea Res Part II Top Stud Oceanogr 57:2064–2075

Schrader H, Swanberg IL, Burckle LH, Grønlien L (1993) Diatoms in recent Atlantic (20°S to 70°N latitude) sediments: abundance patterns and what they mean. Hydrobiologia 269:129–135. https://doi.org/10.1007/BF00028012

Sieracki ME, Verity PG, Stoecker DK (1993) Plankton community response to sequential silicate and nitrate depletion during the 1989 North Atlantic spring bloom. Deep Sea Res Part II Top Stud Oceanogr 40:213–225. https://doi.org/10.1016/0967-0645(93)90014-E

Stemmann L, Hosia A, Youngbluth MJ, Soiland H, Picheral M, Gorsky G (2008) Vertical distribution (0–1000 m) of macrozooplankton, estimated using the underwater video profiler, in different hydrographic regimes along the northern portion of the mid-Atlantic ridge. Deep Sea Res Part II Top Stud Oceanogr 55:94–105. https://doi.org/10.1016/j.dsr2.2007.09.019

Sutton TT, Porteiro FM, Heino M, Byrkjedal I, Langhelle G, Anderson CIH, Horne J, Søiland H, Falkenhaug T, Godø OR, Bergstad OA (2008) Vertical structure, biomass and topographic association of deep-pelagic fishes in relation to a mid-ocean ridge system. Deep Sea Res Part II Top Stud Oceanogr 55:161–184

Sutton TT, Wiebe PH, Madin L, Bucklin A (2010) Diversity and community structure of pelagic fishes to 5000 m depth in the Sargasso Sea. Deep Sea Res Part II Top Stud Oceanogr 57:2220–2233. https://doi.org/10.1016/j.dsr2.2010.09.024

Tréguer P, Nelson DM, Vanbennekom AJ, Demaster DJ, Leynaert A, Quéguiner B (1995) The silica balance in the world ocean: a reestimate. Science 268:375–379

Tréguer P, Bowler C, Moriceau B, Dutkiewicz S, Gehlen M, Aumont O, Bittner L, Dugdale R, Finkel Z, Guidi L, Iudicone D, Jahn O, Lasbleiz M, Levy M, Pondaven P (2018) Influence of diatom diversity on the ocean biological carbon pump. Nat Geosci 11:27–37

van Haren H, Compton TJ (2013) Diel vertical migration in deep sea plankton is finely tuned to latitudinal and seasonal day length. PLoS One 8:e64435. https://doi.org/10.1371/journal.pone.0064435

Vecchione M, Bergstad OA, Byrkjedal I, Falkenhaug R, Gebruk AV, Godø OR, Gislason A, Heino M, Høines AS, Menezes G, Piatkowski U, Priede IG, Skov H, Søiland H, Sutton T, de Lange Wenneck T (2010a) Biodiversity patterns and processes on the Mid-Atlantic Ridge. In: McIntyre A (ed) Life in the world's oceans: diversity, distribution, and abundance. Wiley Blackwell, Oxford, pp 103–121

Vecchione M, Young RE, Piatkowski U (2010b) Cephalopods of the northern Mid-Atlantic Ridge. Mar Biol Res 6:25–52. https://doi.org/10.1080/17451000902810751

Vereshchaka AL, Vinogradov GM (1999) Visual observations of the vertical distribution of plankton throughout the water column above Broken Spur vent field, Mid-Atlantic Ridge. Deep Sea Res Part I Oceanogr Res Pap 46:1615–1632. https://doi.org/10.1016/S0967-0637(99)00018-7

Vereshchaka A, Abyzova G, Lunina A, Musaeva E (2017) The deep-sea zooplankton of the North, Central and South Atlantic: biomass, abundance, diversity. Deep Sea Res Part II Top Stud Oceanogr 137:89–101

Vinogradov GM, Vereshchaka AL (2006) Zooplankton distribution above the Lost City (Atlantis Massif) and Broken Spur hydrothermal fields in the North Atlantic according to the data of visual observations. Oceanology 46:217–227

Vinogradov ME, Vereshchaka AL, Shushkina EA (1996) Vertical structure of the zooplankton community in the oligotrophic areas of the North Atlantic and influence on the hydrothermal vent. Okeanologiya 36:64–71

Vinogradov GM, Vereshchaka AL, Aleinik DL (2003) Zooplankton distribution over hydrothermal fields of the Mid-Atlantic Ridge. Oceanology 43:656–669

Wilson DE, Reeder DAM (2005) Cetacea. In: Mammal species of the world. A taxonomic and geographic reference. 3rd edn. Johns Hopkins University Press, Baltimore, MD

Benthic Organisms

Krzysztof Pabis, Piotr Jóźwiak, and Teresa Radziejewska

1 Meiobenthos

With most research on the hydrothermal vent meiobenthos (benthic meiofauna) having been so far carried out in the Pacific ridge areas (Zeppilli et al. 2018), the Mid-Atlantic Ridge (MAR) meiobenthos appears greatly understudied (Boschen-Rose and Colaço 2021). There have been few dedicated studies (Flint 2007; Zekely et al. 2006; Cuvelier et al. 2014; Sarrazin et al. 2015; Tchesunov 2015; Spedicato et al. 2020), and some additional data can be gleaned from more general overviews of the MAR benthic communities (Gebruk and Mironov 2006).

The MAR meiobenthos has been so far sampled at various known active hydrothermal fields (Menez Gwen, Lucky Strike, Rainbow, Lost City, Broken Spur, TAG, Snake Pit, Logatchev), two of which (Lost City and Broken Spur) are situated within the Polish contract area.

The meiobenthos composition was reported from samples collected, as a rule, with push-corers from a remote-operated vehicle (ROV) (Flint 2007) or a human-operated vehicle (HOV) (*MIR 1* in Tchesunov 2015; *Nautile* in Spedicato et al. 2020); Zekely et al. (2006) collected their samples with a special device called 'Bushmaster Jr.' operated from a Human-Operated Vehicle (HOV). The meiobenthos studied by Sarrazin et al. (2015) was collected with a suction sampler or the ROV's arm grab.

K. Pabis · P. Jóźwiak
Faculty of Biology, Uniwersity of Łódź, Łódź, Poland
e-mail: piotr.jozwiak@biol.uni.lodz.pl

T. Radziejewska (✉)
Institute of Marine and Environmental Sciences, University of Szczecin, Szczecin, Poland
e-mail: teresa.radziejewska@usz.edu.pl

T. Radziejewska et al. (eds.), *The Natural Environment of the Mid-Atlantic Ridge*, GeoPlanet: Earth and Planetary Sciences,
https://doi.org/10.1007/978-3-031-51865-2_5

Generally, the number of samples analysed for the meiobenthos (i.e., the sampling effort) has been low; usually, the researchers had access to 2–12 cores/samples.

The composition of meiobenthic assemblages (Table 1A) shows the presence of a total of 45 taxa. Numerous of them were identified to the species level, although many were reported at a much lower taxonomic resolution (genus, family or order).

All the meiofaunal studies on the MAR show the meiobenthos assemblages to consist of the Nematoda, Gastrotricha, Polychaeta, Halacaroidea, Ostracoda, Copepoda (Calanoida, Cyclopoida, Harpacticoida, Poecilostomatoida, Siphonostomatoida), Tantulocarida and Bivalvia (Table 1A). The hydrothermal fields within the Polish contract area (Lost City and Broken Spur) have proven very impoverished in this respect, and also differing in terms of their taxonomic richness: while no non-nematode taxon has been reported from the Lost City, the Broken Spur turned out to support 5 meiofaunal taxa, all of them representing copepods. Noteworthy is the presence there of 3 species of the copepod siphonostomatoid family Dirivultidae, regarded as endemic to the hydrothermal vents worldwide and particularly well adapted to life in the vicinity of the hot venting fluid (Zeppilli et al. 2018).

As shown in Fig. 1, the similarity in faunistic composition between meiobenthic assemblages reported so far from different vent fields on the MAR is quite low, the assemblages for which data are known from the Broken Spur and Lost City being, interestingly, more similar to one another (and to that at the Rainbow) than to the remaining fields. It has to be remembered, however, that the similarity analysis shown in Fig. 1 has been carried out on a very limited amount of data.

All the MAR studies report the meiobenthos assemblages to be dominated by nematodes, copepods (notably harpacticoids) being the second most abundant taxon; it was only at a non-vent site at the TAG field sampled by Flint (2007) that the proportion between the two higher taxa was reversed. On account of the nematode dominance, most of the meiofaunal studies on the MAR have been, in fact, focussing on nematodes. This is evident from the list of meiofaunal taxa reported from the MAR hydrothermal vent fields provided by the relevant publications. In fact, the list of nematode taxa (mostly genera, but also species and families) is quite long (a total of 74), considering the low number of samples analysed so far (Table 1B).

The hydrothermal fields within the Polish contract area (Lost City and Broken Spur) have been reported to differ in terms of their nematode taxonomic richness (Table 1B). While 13 nematode taxa have been reported from the Lost City, as few as 3 have been found at the Broken Spur. Interestingly, the Lost City nematode assemblage features the genus *Oncholaimus*, reported as very abundant at most active vent sites (Tchesunov 2015; Zeppilli et al. 2018). The genus members specialise in feeding on free-living chemoautotrophic microorganisms and harbour an epibiotic microbial community dominated by the Gammaproteobacteria lineage, including vent symbionts (Zeppilli et al. 2018).

Recently, Washburn et al. (2023) reported on benthic community responses to the first massive polymetallic sulphide mining tests in Japanese waters. Among the meiobenthos, they explored for specific nematode taxa as indicators of disturbance

Table 1 Meiofaunal taxa reported from the hydrothermal vent fields in the Polish contract area (Lost City, Broken Spur) and elsewhere on MAR; (A) non-nematode taxa; (B) nematode taxa (data from Gebruk and Mironov 2006; Flint 2007; Tchesunov 2015; Zeppilli et al. 2015; Sarrazin et al. 2015; Spedicato et al. 2020); +, taxon recorded; –, taxon not recorded

Taxon		Lost City	Broken Spur	Other MAR fields
A				
Gastrotricha	Gastrotricha indet.	–	–	+
Polychaeta	Polychaeta indet.	–	–	+
Halacaroidea	*Agauopsis auzendei*	–	–	+
	Bathyalacarus sp.	–	–	+
	Halacarellus alvinus	–	–	+
	Halacarus prolongatus	–	–	+
	Copidognatus alvinus	–	–	+
	C. nautilei	–	–	+
Tantulocarida	Tantulocarida indet.	–	–	+
Copepoda Siphonostomatoida	*Aphotopontius atlanteus*	–	–	+
	A. temperatus	–	–	+
	A. forcipatus	–	–	+
	Rimipontius mediospinifer	–	+	+
	Stygiopontius rimivagus	–	–	+
	S. bulgisetiger	–	–	+
	S. cladarus	–	+	+
	S. latulus	–	–	+
	S. mirus	–	–	+
	S. pectinatus	–	+	+
	S. regius	–	–	+
	S. serratus	–	–	+
	S. teres	–	–	+
	Dirivultidae indet.	–	–	+
Copepoda Harpacticoida	*Bathylaophonte azorica*	–	–	+
	Microsetella norvegica	–	+	–
	Aegisthidae indet.	–	–	+
	Ectinosomatidae indet.	–	–	+
	Miraciidae indet.	–	–	+
	Pseudotachidiidae indet.	–	–	+
	Tegastidae indet.	–	–	+
	Tisbidae indet.	–	–	+
Copepoda Cyclopoida	*Heptnerina confusa*	–	–	+
	Oncaea sp.	–	+	–
	Cyclopinidae indet.	–	–	+
	Oithonidae indet.	–	–	+
	Lubbockiidae	–	–	+
Copepoda Spinocalanoida	Spinocalanoida indet.	–	–	+

(continued)

Table 1 (continued)

Taxon		Lost City	Broken Spur	Other MAR fields
Ostracoda	*Bathyconchoecia paulula*	–	–	+
	Bairdia sp.	–	–	+
	Bythocypris sp.	–	–	+
	Krithe sp.	–	–	+
	Propontocypris sp.	–	–	+
	Poseidonamicus sp.	–	–	+
	Ostracoda indet.	–	–	+
Bivalvia	Bivalvia indet.	–	–	+

B

Taxon	Lost City	Broken Spur	Other MAR fields
Acantholaimus	–	–	+
Acanthopharynx	–	–	+
Actinonema	+	–	–
Anoplostoma	–	–	+
Anticoma	–	–	+
Anticomidae indet.	+	+	+
Amphimonhystrella	–	–	+
Aponema	–	–	+
Araeolaimus	–	–	+
Benthimermitidae indet.	–	–	+
Calomicrolaimus	–	–	+
Camacolaimus	–	–	+
Capsula	–	–	+
Cephalochaetosoma	–	–	+
Chromadorella	–	–	+
Chromadorina	–	–	+
Chromadorita	–	–	+
Cobbia	–	–	+
Cyatholaimus	–	–	+
Daptonema	+	–	+
Desmodora	+	–	+
Desmoscolex	–	–	+
Dinetia	–	–	+
Diplopeltidae indet.	–	–	+
Diplopeltoides	–	–	+
Dracogallus	–	–	+
Draconema	–	–	+
Eleutherolaimus	–	–	+
Enchelidiidae indet.	+	–	–
Enoplus	+	–	–
Enoplidae indet.	–	–	+

Table 1 (continued)

B			
Taxon	Lost City	Broken Spur	Other MAR fields
Epsilonema	–	–	+
Epsilonematidae indet.	–	–	+
Euchromadora	–	–	+
Eurystomina	–	–	+
Halalaimus	–	–	+
Halomonhystera	–	–	+
Halichoanalaimus	–	–	+
Leptolaimus	–	–	+
Leptolaimidae indet.	–	–	+
Megadesmolaimus	–	–	+
Metacomesoma	–	–	+
Metacylicolaimus	+	–	–
Metachromadora	–	–	+
Metadesmolaimus	–	–	+
Metalinhomoeus	–	–	+
Metepsilonema	–	–	+
Microlaimus	–	–	+
Molgolaimus	–	–	+
Monhysteridae indet.	–	–	+
Oncholaimus	+	–	+
Oncholaimidae indet.	–	–	+
Oxystomina	–	–	+
Palinchomeous	+	–	+
Paracanthonchus	–	–	+
Paracanthonchus olgae	+	+	+
Parapinnanema	–	–	+
Prochaetosoma ventriverruca	+	–	+
Prochromadora helenae	+	+	+
Prooncholaimus	–	–	+
Pselionema	–	–	+
Quadricoma	–	–	+
Retrotheristus	–	–	+
Rhabdocoma	–	–	+
Sabatieria	–	–	+
Sphaerolaimus	–	–	+
Synonchus	–	–	+
Syringolaimus	+	–	+
Terschellingia	–	–	+
Thalassolaimus	–	–	+
Thalassomonhystera	–	–	+
Theristus	–	–	+
Tricoma	–	–	+
Viscosia	–	–	+

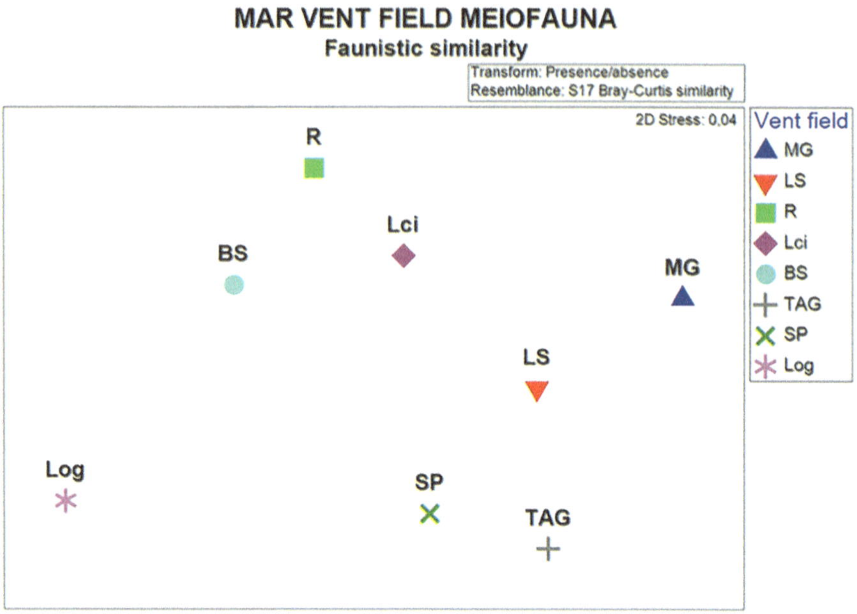

Fig. 1 MDS plot of similarity/dissimilarity between meiobenthos faunistic composition at different MAR vent fields (presence-absence transformation; Bray-Curtis similarity); Vent field name symbols: R, Rainbow; BS, Broken Spur; Lci, Lost City; MG, Menez Gwen; LS, Lucky Strike; SP, Snake Pit; TAG, Trans-Atlantic Geotraverse; Log, Logatchev (source: Radziejewska et al. 2022)

from mining activities. They found high abundances of *Monhystera*, *Daptonema* and *Desmodora* nematodes as indicative of areas impacted by mining. In contrast, the genera *Halalaimus*, *Molgolaimus*, *Sphaerolaimus*, *Oxystomina*, *Diplopeltoides*, *Halichoanolaimus*, *Pselionema* and *Leptolaimus* to be indicators of the lack of impact. Representatives of *Desmoscolex* and *Actinonema* appeared to be equally abundant in both impacted and non-disturbed locations.

At hydrothermal vent sites worldwide, the meiobenthos has been found to be associated with various habitats, from inactive bare basalts (Gollner et al. 2010) and sediments (Vanreusel et al. 1997) to diffuse vent flow areas colonised by macrofaunal assemblages (Zekely et al. 2006; Sarrazin et al. 2015). When working in the Pacific (Fiji-Lau area), Vanreusel et al. (1997) found none of the nematode species found in the hydrothermal sediments to occur in the surrounding abyssal areas. No such comparisons have been conducted on the MAR.

Meiofaunal (nematode) colonisation of artificially deployed, organic (wood and bone) and inorganic (slate), substrata at the MAR's Lucky Strike hydrothermal vent site was addressed by a French research group (Cuvelier et al. 2014; Zeppilli et al. 2015). A fast colonisation as well as a clear preference with respect to the substratum type, depending on the distance from the vent, was reported, particularly evident during the first 9 months of the study. Such experiments are important as they provide data with which to assess the community's ability to recover from a disturbance, and confirm the utility of the meiobenthos in studies on post-disturbance recovery.

2 Macrobenthos

Macrofaunal benthic communities of the Mid-Atlantic Ridge (MAR) have been studied less extensively than the assemblages of megafaunal invertebrates (see below). More comprehensive data sets concern the larger epibenthic macrofauna, such as echinoderms and all the other organisms visible in seafloor imagery, including the temporary macrofauna, e.g., the *Rimicaris* shrimps. Therefore, data on macrofaunal distribution patterns have been, at least in part, acquired as a 'side effect' of studies focussing on the megafaunal communities, for which reason identification has been often limited to higher taxonomic units (genera or families). The information provided in publications makes it occasionally difficult to conclude whether the analysis concern the macro- or the megafaunal organisms (e.g., Desbruyères et al. 2000; Kelley et al. 2005; Alt et al. 2013, 2019; Bell et al. 2016).

Knowledge on the most important and diverse macrofaunal taxa, such as polychaetes, peracarid crustaceans, bivalves and gastropods is very scant, and there have been only a few quantitative assessments of the standing stock and diversity. The limitation is due to constraints associated with the so-called physical sampling that requires the use of dredges, epibenthic sledge or various grabs and corers, as well as to the need of taxonomic expertise. A particularly wide knowledge gap concerns the fauna associated with soft sediments (Bergstad et al. 2008; Gebruk et al. 2010; Shields and Blanco-Perez 2013; Priede et al. 2013), although this habitat is a very important component of the MAR region (Niedzielski et al. 2013).

There are almost no data on the biology and habitat requirements of most of the macrofaunal taxa. Some case studies revealed certain very interesting interspecific associations, such as those between polynoid polychaetes and holothurians or bivalves (Shields et al. 2013; Bebianno et al. 2018), demonstrating a unique nature of biological interactions, particularly in the vent systems. Other authors elucidated differences in the composition, abundance, and diversity of macrofaunal assemblages, shaped by key environmental factors (temperature, total dissolved iron, sulphide and copper, pH), but on a very small scale (less than 1 m^2) (Sarrazin et al. 2015). Studies of the reproductive behaviour, physiology, and development have also been infrequent, a particular attention being paid to the vent fauna (e.g., Blake and Van Dover 2005; Riou et al. 2010; Marticorena et al. 2020).

The available species lists are often scattered among various publications, including short taxonomic notes, and there is a limited information on distribution patterns and factors controlling the diversity, species composition, and abundance of the communities (e.g., Bellan-Santini and Thurston 1996; Gebruk et al. 2010; Goroslavskaya and Galkin 2011; Dilman 2013; Shields and Blanco-Perez 2013; Kongsrud et al. 2013).

The scale of the undescribed biodiversity is most probably huge. The study of polychaete fauna collected with a megacorer (Shields and Blanco-Perez 2013) resulted in finding 133 species in just 11 samples (only 0.54 m^2 of the total surface sampled). The mean abundance reached 1285 ± 464 ind./m^2, while the mean biomass was 19.60 ± 713.27 mg/m^2. The fauna was dominated by spionids (mainly *Aurospio dibranchiata* and *Prionospio* sp.), syllids, and cirratulids. The communities were substantially (in 70%) dominated by surface deposit feeders. Most of the species collected were singletons, therefore a high level of rarity can be assumed, although the final conclusion is difficult without any higher sampling effort. Many of the species collected are also new to science, although most of them are still awaiting the formal description. Similar results were reported for scavenging amphipods, with about 50% of the species being previously unknown (Horton et al. 2013).

It is also difficult to properly assess the level of endemism in the most important invertebrate groups (cf. Chapter 6), although the few available taxonomic studies showed the presence of new species on the MAR, including in vent habitats (e.g., Tyler et al. 1995; Sigvaldadottir and Desbruyères 2003; Myers and Cunha 2004; Cunha and Wilson 2006; Larsen et al. 2006; Bellan-Santini 2007; Corbera et al. 2008; Budaeva 2012), and there is a great need for an increased taxonomic effort in the whole region. Many new species can be expected among small peracarids with limited dispersal potential, such as tanaidaceans and isopods, which are a very important component of the deep-sea communities worldwide (e.g., Błażewicz-Paszkowycz et al. 2015; Pabis et al. 2015; Riehl et al. 2018). Molecular studies focussed on the analysis of species composition and diversity are almost absent, whereas such studies might be crucial for an appropriate future assessment of impacts produced by mining activities (Klunder et al. 2020).

Analyses of spatial distribution of some taxa showed a certain division between the north-east and south-east sites in the case of polychaetes (Shields and Blanco-Perez 2013), whereas scavenging amphipods were shown to be more widely distributed, most probably as a result of their higher mobility and feeding mode (Horton et al. 2013). Generally, the abundance and biomass of the macrofauna seems to be at a level similar to that in other deep-sea sites (Priede et al. 2013), although some scavengers such as amphipods may form temporary aggregations of even 40,000 individuals, food availability allowing (Horton et al. 2013). Jones et al. (2013) showed also that enteropneusts can be important bioturbators at bathyal depths of the MAR.

As summarised by Kelley et al. (2005) for MAR vent sites, the number of species per site may vary between 30 and 50, while for eight major MAR vent fields about 100 species were recorded at each location. The fauna of the hydrothermal vent fields located in the Polish contract area has been studied intermittently only, and the available data likely provide a highly incomplete picture of its diversity, abundance, and biomass. The data on the Lost City macrofauna are very scant. Nevertheless, the available reports show the presence of at least 65 species (collected with a suction sampler), including gastropods, amphipods (e.g., *Bouvierella curtirama, Primno evansis*), polychetes, euphausiids, and ostracods. The level of endemism has been estimated at about 58% (Kelley et al. 2005, 2007), although recent molecular studies demonstrated that the problems of endemism, cryptic diversity and connectivity along the MAR are very complex (cf. Chapter 6) and require further studies (Teixeira et al. 2013; Yahagi et al. 2019), which are important for appropriate conservation strategies and/or management plans associated with any type of human activities that may influence seabed communities (Hilário et al. 2015). Information on the Broken Spur vent field macrobenthos is also fairly scant; that fauna includes, for example, the bivalve *Bathymodiolus puteoserpentis*, the ampharetid polychaete *Amathys lutzi* and the ophiuroid *Ophioctenella acies* (Rybakova and Galkin 2014) as well as temporarily macrofaunal shrimps or polychaetes that can attain much larger sizes during their development and can thus be classified as the megafauna (Desbruyères et al. 2000). A list of non-meiobenthic taxa reported so far from the Polish contract area vent fields (Lost City and Broken Spur) is provided in Table 2. (cf. Radziejewska et al. 2022).

A multivariate analysis of faunistic similarity between the macrobenthos of individual vent sites (Boschen-Rose and Colaço 2021) showed some similarity between

Table 2 Non-meiobenthic taxa reported from the Polish exploration contract area (a compilation of data from publications cited)

Taxon			Lost City	Broken Spur
Cnidaria	Hydrozoa	*Candelabrum* sp.	–	+
	Actiniaria	*Maractis rimicarivora*	–	+
	Scleractinia	*Demosphyllum pertusum*	+	–
Annelida	Polychaeta	*Amphisamytha lutzi*	–	+
		Archinome sp.	–	+
		Ophiotrocha sp.	+	–
		Branchiopolynoe aff. *seepensis*	–	+
		Prionospio sp.	–	+
Mollusca	Gastropoda	*Lirapex* sp.	+	–
		Lurifax vitreus	+	–
		Peltospira smaragdina	–	+
		Phymorchynchus carinatus	–	+
		Phymorchynchus moskalevi	–	+
		Phymorhynchus ovatus	–	+
		Phymorchynchus sp.	+	–
		Pseudorimula midatlantica	–	+
		Puncturella sp.	+	–
		Shinkailepas briandi	–	+
		Emarginula sp.	+	–
		Lepetodrilus atlanticus	–	+
	Bivalvia	*Asperarca nodulosa*	+	–
		Bathymodiolus azoricus	+	+
		Bathymodiolus puteoserpentis	–	+
		Thyasira sp.	+	–

(continued)

Table 2 (continued)

Taxon			Lost City	Broken Spur
Arthropoda	Cirripedia	*Metacerrucca inermis*	+	–
		Poecilasma aurantia	+	–
	Amphipoda	*Bouvierella* aff. *curtirama*	+	–
		Hirondella brevicaudata	–	+
		Paraphronima sp.	–	+
		Primno evansi	–	+
		Streetsia sp.	–	+
		Dulichiopsis dianae	–	+
	Decapoda	*Alvinocaris markensis*	–	+
		Chorocaris chacei	–	+
		Mirocaris fortunata	–	+
		Rimicaris exoculata	–	+
		Keldyshicaris vavilovi	–	+
		Bathynectes maravigna	+	–
		Segonzacia mesatlantica	–	+
		Munida sp.	+	–
		Munidopsis exuta	–	+
Echinodermata	Ophiuroidea	*Ophioctenella acies*	–	+
	Echinoidea	*Araeosoma fenestratum*	+	–
		Cidaris sp.	+	–

the Broken Spur fauna and benthic communities sampled at other vent fields, including the TAG, Snake Pit, Logatchev and Ashadze-1. Based on the available data (Gebruk and Mironov 2006), the Lost City fauna showed a low similarity to the macrobenthos at other vent fields.

The available data demonstrate the MAR macrofauna to be a diverse and abundant component of a very complex system that needs further, much more comprehensive research based on a higher sampling effort, accompanied by the analysis of various physico-chemical factors. At the moment, we have probably managed to get a glimpse on a tip of an iceberg.

3 Megabenthos

The general knowledge about benthic megafaunal (megabenthic) communities of the Mid-Atlantic Ridge (MAR) is still insufficient, although—considering the substantial undersampling of the deep-sea benthic communities worldwide (Menegotto and Rangel 2018) and large gaps in knowledge on benthic deep-sea faunas, even at the basic level of species richness (Costello and Chaudary 2017)—the area is relatively well-studied (Perez et al. 2012; Priede et al. 2013; Thaler and Amon 2020 and references therein).

The MAR has been studied in the framework of several international projects, for example the Mid-Atlantic Ridge Ecosystem Project (MAR-ECO) which resulted in papers about the distribution patterns and taxonomy of selected benthic megafaunal invertebrates, such as sponges (Hestetun et al. 2015; Cardenas and Rapp 2015), corals (Buhl-Mortensen et al. 2008; Molodtsova et al. 2008; Morris et al. 2012; Braga-Henriques et al. 2013) or echinoderms (Martynov and Litvinova 2008; Gebruk 2008; Dilman 2008, 2013; Rogacheva et al. 2013). The communities are dominated by surface deposit feeders and suspension feeders (Alt et al. 2019).

Differences in the distribution of megafauna on a smaller, local scale might be associated with microhabitat variation (e.g., the type of substratum, the presence of hard bottom) and food availability (the quantity and quality of organic matter, effects of currents), which may result in temporary aggregations of motile species (Morris et al. 2012; Alt et al. 2019). On the other hand, considerable spatial variability is encountered even in communities of sessile organisms, despite the absence of differences in the substratum type (Bell et al. 2016). Some of the studies revealed also a depth-related zonation (Gebruk and Krylova 2013; Molodtsova et al. 2017). The role of sessile megafaunal communities must also be viewed through their importance as ecosystems engineers that may shape the overall biodiversity by increasing the habitat complexity and providing microhabitats for other invertebrates. Generally, recovery of such sessile communities after disturbances is known to take a long time, resulting in a substantial impoverishment of the associated fauna of smaller organisms (Buhl-Mortensen et al. 2008; Alt et al. 2019).

Research on the megafaunal biomass demonstrated that the observed values were similar to those reported from other deep-sea locations (Priede et al. 2013).

The high habitat heterogeneity, especially along steep slopes, results in mosaic patches of different taxa, although it is worth mentioning that distribution patterns may change horizontally along the MAR (Gebruk et al. 2010; Alt et al. 2013, 2019; Rogacheva et al. 2013; Bell et al. 2016; Afonso et al. 2020).

The hydrothermal vents are particularly important components of the MAR ecosystem, with the Lost City hydrothermal field (in the Polish contract area) being a unique and the most striking example. Discovered only 20 years ago, the field is not only a unique habitat (cf. Chapter 3), but also a biodiversity hotspot with a high number of endemic taxa. A multivariate analysis of faunistic similarity between megafaunal assemblages at the eight MAR hydrothermal vent fields (Fig. 2), based on data provided by Gebruk and Mironov (2006), produced a very interesting pattern, with the Lost City megafauna being the least similar to the assemblages recorded at the remaining fields. Those assemblages, in turn, are separated into a northern (Rainbow, Lucky Strike, Menez Gwen) and southern (Broken Spur, TAG, Logatchev, and Snake Pit) component. Future research should show whether this pattern is a persistent one.

Generally, the megafauna in vent habitats is dominated by motile invertebrates, including large crabs of the family Geryonidae. Non-venting habitats located close to active areas are inhabited by corals (e.g., *Lophelia* sp., *Desmophyllum* sp.), but also by motile invertebrates, including echinoderms and crabs of the family Galatheidae. Despite the relatively low biomass, the Lost City ecosystem likely

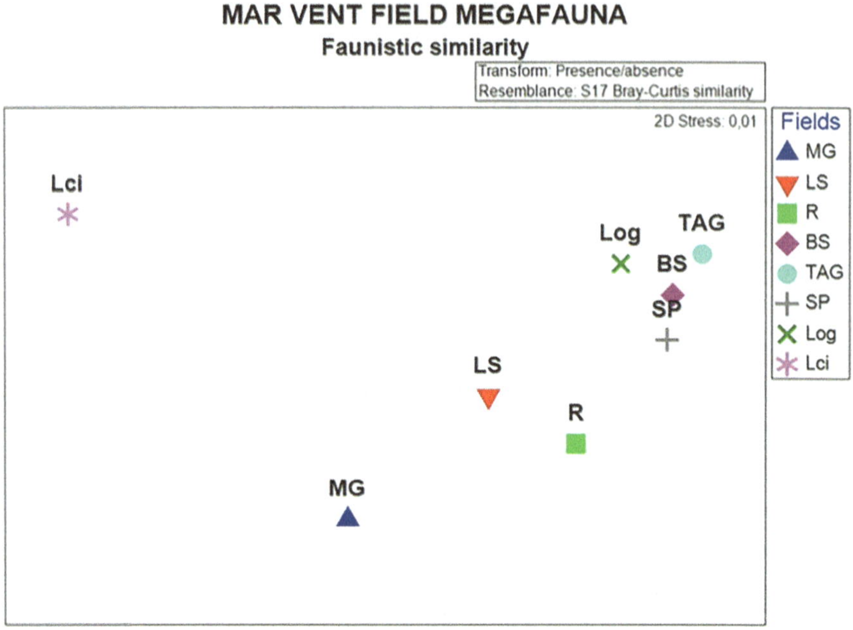

Fig. 2 MDS plot of similarity/dissimilarity between megabenthic faunistic composition at different MAR vent fields (presence-absence transformation; Bray-Curtis similarity); Vent field name symbols: R, Rainbow; BS, Broken Spur; Lci, Lost City; MG, Menez Gwen; LS, Lucky Strike; SP, Snake Pit; TAG, Trans-Atlantic Geotraverse; Log, Logatchev (based on data extracted from publications cited)

supports the highest diversity among all the MAR hydrothermal vent fields surveyed (Kelley et al. 2005, 2007; Johnson 2019). The megafauna at the Broken Spur, another hydrothermal field in the Polish contract area, is dominated by shrimps (e.g., *Rimicaris exoculata*, particularly on chimney walls), crabs (*Segonzacia mesatlantica*), polychaetes (e.g., *Chaetopterus*) and actiniarians (Desbruyères et al. 2000). Some studies suggest that, even in the areas away from the hydrothermal vent ecosystems, the contribution of endemic species may reach several per cent in some groups (e.g., Molodtsova et al. 2008; Tabachnick and Collins 2008; Braga-Henriques et al. 2013) (see below).

It has to be borne in mind that the general knowledge on the megafaunal communities on the MAR, and the Polish contract area in particular, is based on only a relatively low number of published reports. Therefore, any generalisations and more comprehensive conclusions regarding the distribution patterns and factors influencing the megabenthic communities are very tentative and need verification by further studies based on a higher sampling effort. Moreover, methodological constraints of megafaunal studies (e.g., commonly conducted observations from ROVs) produce low-resolution taxonomic assessments leading to underestimation of the diversity (e.g., Alt et al. 2019). However, the technical advances in imaging technology (e.g.,

Dumke et al. 2018) and annotation software (Gomes-Pereira et al. 2016) have allowed a progress to be made in this respect. Also, methods are at present in use with which to estimate the megafaunal biomass from underwater imagery (Benoist et al. 2019). Nevertheless, it is still very difficult to identify factors shaping spatial patterns of the megabenthos and to predict possible responses of the communities to anthropogenic impacts (Dunn et al. 2018).

4 Nektobenthos

Since there are methodological difficulties in examining the MAR area, and because the majority of the studies have been based on video recordings, data on the nektobenthos are scarce and often limited to higher taxa (e.g., Biscoito et al. 2006; Felley et al. 2008).

The nektobenthos on the MAR is represented mainly by the typical North Atlantic lower-bathyal species known from the ocean margins, and the endemicity of the MAR species is most probably not high, e.g., non-existent in both fishes and cephalopods (Priede et al. 2013). A comparison of the MAR fish fauna with that of the North Atlantic margins showed stronger similarities between the MAR and the eastern, rather than the western, Atlantic. Moreover, based on their fish data, Priede et al. (2013) pointed out that the biodiversity on the MAR is higher than that in the adjacent abyssal plain.

Shallower areas of the MAR are characterized by a more diverse fauna as well as a higher abundance and biomass than deeper bottom areas (Felley et al. 2008). Similar findings were reported by Bergstad et al. (2008) for demersal fishes, and by Vecchione et al. (2010) for cephalopods. The non-vent nektobenthos is often widely distributed and shows no clear habitat preferences (Horton et al. 2013).

The non-vent nektobenthic fish fauna of the MAR area (cf. Chapter 4) is represented by, e.g., the Macrouridae (including *Coryphaenoides brevibarbis* and *C. armatus*), Halosauridae (e.g., *Halosauropsis macrochir* and *Aldrovandia* sp.), Bathysauridae (e.g., *Bathysaurus ferox*), Notacanthidae (e.g., *Polyacanthonotus challengeri*) and Moridae (e.g., *Antimora rostrata*), all recorded from the Charlie-Gibbs Fracture Zone (CGFZ) (Felley et al. 2008; Linley et al. 2013). Cousins et al. (2013), who used a baited-camera lander, list 19 species, with the dominant *C. armatus* and *A. rostrata*.

Barathronus multidens (Aphyonidae), *Ipnops murray* (Ipnopidae) and *Holtbyrnia anomala* (Platytroctidae) were collected between 16°40′ and 17°14′N (Molodtsova et al. 2017). A transect from Iceland to the Azores (36 stations) brought records of *Eurypharynx pelecanoides*, *Bathylagus euryops*, *Sigmops bathyphilum*, *Scopeloberyx robustus*, *Serrivomer beanii* and *Borostomias antarcticus* (Sutton et al. 2008).

The nektobenthic invertebrates on the MAR feature primarily two groups: decapod crustaceans and cephalopods. The decapod fauna in non-vent MAR habitats includes penaeid shrimps (Penaeidea) and the squat lobster (*Munidopsis* sp.)

reported from the Charlie-Gibbs Fracture Zone (Felley et al. 2008). Horton et al. (2013) listed 39 species of necrophagous amphipods from samples taken north and south of the Charlie-Gibbs Fracture Zone and east and west from the ridge. A total of 31 species on their list represent the superfamily Lysianassoidea, with the most abundant *Abyssorchomene abyssorum*. Vecchione et al. (2010) provided a list of 34 cephalopod species found on the MAR, collected with a bottom trawl north of the Azores and in areas just north and south of the Charlie-Gibbs Fracture Zone. Because of the sampling gear used, those species might be considered as belonging to the nektobenthos, but the authors quoted considered some of them as pelagic forms that contaminated the sample during gear deployment. Fifteen of those 34 species were collected only with the benthic trawl. Among them, 14 species are considered to represent the true bottom fauna, with *Mastigoteuthis agassizii*, *Gonatus steenstrupi* and *Stauroteuthis syrtensis* as the dominant species.

The nektobenthic fish fauna of the vent habitats has been seldom studied. There are, however, some data from the Broken Spur hydrothermal field in the Polish contract area where two vent-associated fish species: *Haptenchelys texis* (Synaphobranchidae) and *Pachycara thermophilum* (Zoarcidae) have been recorded (Parin 1995). Some data from other fields do exist as well. For example, Desbruyères et al. (2006) listed four species (*Gaidropsaurus* sp., *Ilyophis saldanhai*, *Pachycara saldanhai* and *P. termophilum*) as typical of MAR vent fields. *P. thermophilum* (Zoarcidae) was common at the Logatchev field as well (Gebruk et al. 2000). Biscoito et al. (2006) listed 19 species of demersal fishes from the Saldanha vent field. The MAR decapod vent fauna is represented by, e.g., shrimps of the genus *Nematocarcinus* found at the Saldanha vent field (Biscoito et al. 2006) or *Mirocaris fortunata* and *Rimicaris exoculata,* both recorded also at the Broken Spur (Desbruyères et al. 2000, 2001) and often referred to as the megafauna. Single records of the cephalopods *Cirroteuthis* sp. and *Grimpoteuthis* sp. were provided by Biscoito et al. (2006) from the Saldanha vent field.

References

Afonso P, Fontes J, Giacomello E, Magalhães MC, Martins HR, Morato T, Neves V, Prieto R, Santos RS, Silva M, Vandeperre F (2020) The Azores: a mid-Atlantic hotspot for marine megafauna research and conservation. Front Mar Sci 6:826. https://doi.org/10.3389/fmars.2019.00826

Alt CHS, Kremenetskaia A, Boorman B, Hughes JA, Billett DSM, Gooday AJ, Jones DOB (2013) Trawled megafaunal invertebrate assemblages from bathyal depth of the Mid-Atlantic Ridge (48°–54°N). Deep Sea Res Part II Top Stud Oceanogr 98:326–340

Alt CHS, Kremenetskaia A, Gebruk AV, Gooday AJ, Jones DOB (2019) Bathyal benthic megafauna from the Mid-Atlantic Ridge in the region of the Charlie-Gibbs Fracture Zone based on remotely operated vehicle observations. Deep Sea Res Part I Oceanogr Res Pap 145:1–12

Bebianno MJ, Cardoso C, Gomes T, Blasco J, Santos RS, Colaço A (2018) Metal interactions between the polychaete *Branchipolynoe seepensis* and the mussel *Bathymodiolus azoricus* from Mid-Atlantic-Ridge hydrothermal vent fields. Mar Environ Res 135:70–81

Bell JB, Alt CHS, Jones DOB (2016) Benthic megafauna on steep slopes at the northern Mid-Atlantic Ridge. Mar Ecol 37:1290–1302

Bellan-Santini D (2007) New amphipods of hydrothermal vent environments on the Mid-Atlantic Ridge, Azores Triple Junction zone. J Nat Hist 41:567–596
Bellan-Santini D, Thurston MH (1996) Amphipoda of the hydrothermal vents along the Mid-Atlantic Ridge. J Nat Hist 30:685–702
Benoist NMA, Bett BJ, Morris KJ, Ruhl HA (2019) A generalised volumetric method to estimate the biomass of photographically surveyed benthic megafauna. Prog Oceanogr 178:102188
Bergstad OA, Falkenhaug T, Astthorsson OS, Byrkjedal I, Gebruk AV, Piatkowski U, Priede IG, Santos RS, Vecchione M, Lorance P, Gordon JDM (2008) Towards improved understanding of the diversity and abundance patterns of the mid-ocean ridge macro- and megafauna. Deep Sea Res Part II Top Stud Oceanogr 55:1–5
Biscoito M, Almeida AJ, Segonzac M (2006) Preliminary biological characterization of the Saldanha hydrothermal field at the Mid-Atlantic Ridge (36°34′N, 32°26′W, 2200 m). Cah Biol Mar 47:421–427
Blake EA, Van Dover CL (2005) The reproductive biology of *Amathys lutzi*, an ampharetid polychaete from hydrothermalvents on the Mid-Atlantic Ridge. Invertebr Biol 124:254–264
Błażewicz-Paszkowycz M, Pabis K, Jóźwiak P (2015) Tanaidacean fauna of the Kuril-Kamchatka Trench and adjacent abyssal plain – abundance, diversity and rare species. Deep Sea Res Part II Top Stud Oceanogr 111:325–332
Boschen-Rose RE, Colaço A (2021) Northern Mid-Atlantic Ridge hydrothermal habitats: a systematic review of knowledge status for environmental management. Front Mar Sci 8:657358
Braga-Henriques A, Porteiro FM, Ribeiro PA, de Matos V, Sampaio I, Ocaña O, Santos RS (2013) Diversity, distribution and spatial structure of the cold-water coral fauna of the Azores (NE Atlantic). Biogeosciences 10:4009–4036
Budaeva N (2012) *Leptoecia midatlantica*, a new species of the deep-sea quill-worms (Polychaeta: Onuphidae: Hyalinoeciinae) from the Mid-Atlantic Ridge. Zootaxa 3176:45–60
Buhl-Mortensen P, Buhl-Mortensen L, Gebruk AV, Krylova EM (2008) Occurrence of deep-water corals on the Mid-Atlantic Ridge based on MAR-ECO data. Deep Sea Res Part II Top Stud Oceanogr 55:142–152
Cardenas P, Rapp T (2015) Demosponges from the northern Mid-Atlantic Ridge shed more light on the diversity and biogeography of North Atlantic deep-sea sponges. J Mar Biol Assoc U K 95:1475–1516
Corbera J, Segonzac M, Cunha MR (2008) A new deep-sea genus of Nannastacidae (Crustacea, Cumacea) from the Lucky Strike hydrothermal vent field (Azores Triple Junction, Mid-Atlantic Ridge). Mar Biol Res 4:180–192
Costello MJ, Chaudary C (2017) Marine biodiversity, biogeography, deep-sea gradients, and conservation. Curr Biol 27:511–527
Cousins NJ, Linley TD, Jamieson AJ, Bagley PM, Blades H, Box T, Chambers R, Ford A, Shields MA, Priede IG (2013) Bathyal demersal fishes of Charlie-Gibbs Fracture Zone region (49–54°N) of the Mid-Atlantic Ridge: II. Baited camera lander observations. Deep Sea Res Part II Top Stud Oceanogr 98:397–406
Cunha MR, Wilson GDF (2006) The North Atlantic genus *Heteromesus* (Crustacea: isopoda: Asellota: Ischnomesidae). Zootaxa 1192:3–76
Cuvelier D, Beesau J, Ivanenko VN, Zeppilli D, Sarradin PM, Sarrazin J (2014) First insights into macro- and meiofaunal colonisation patterns on paired wood/slate substrata at Atlantic deep-sea hydrothermal vents. Deep Sea Res Part I Oceanogr Res Pap 87:70–81
Desbruyères D, Almeida A, Biscoito M, Comtet T, Khripounoff A, Le Bris N, Sarradin PM, Segonzac M (2000) A review of the distribution of hydrothermal vent communities along the northern Mid-Atlantic Ridge: dispersal vs. environmental controls. Hydrobiologia 440:201–216
Desbruyères D, Biscoito M, Caprais J-C, Colaço A, Comtet T, Crassous P, Fouquet Y, Khripounoff A, Le Bris N, Olu K, Riso R, Sarradin P-M, Segonzac M, Vangriesheim A (2001) Variations in deep-sea hydrothermal vent communities on the Mid-Atlantic Ridge near the Azores Plateau. Deep Sea Res Part I Oceanogr Res Pap 48:1325–1346

Desbruyères D, Segonzac M, Bright M (2006) Handbook of deep-sea hydrothermal vent fauna. Denisia, Linz-Dornach

Dilman A (2008) Asteroid fauna of the northern Mid-Atlantic Ridge with description of a new species *Hymenasterides mironovi* sp. nov. Mar Biol Res 4:131–151

Dilman A (2013) Asteroid fauna of the northern Mid-Atlantic Ridge: additional records. Mar Biol Res 9:563–586

Dumke I, Purser A, Marcon Y, Nornes SM, Johnsen G, Ludvigsen M, Søreide F (2018) Underwater hyperspectral imaging as an *in situ* taxonomic tool for deep-sea megafauna. Sci Rep 8:12860

Dunn DC, Van Dover CL, Etter RJ, Smith CR, Levin LA, Morato T, Colaço A, Dale AC, Gebruk AV, Gjerde KM, Halpin PN, Howell KL, Johnson D, Perez JAA, Ribeiro MC, Stuckas H, Weaver P, SEMPIA Workshop Participants (2018) A strategy for the conservation of biodiversity on mid-ocean ridges from deep-sea mining. Sci Adv 4:eaar4313

Felley JD, Vecchione M, Wilson RR (2008) Small-scale distribution of deep-sea demersal nekton and other megafauna in the Charlie-Gibbs Fracture Zone of the Mid-Atlantic Ridge. Deep Sea Res Part II Top Stud Oceanogr 55:153–160. https://doi.org/10.1016/j.dsr2.2007.09.021

Flint HC (2007) Diversity of meiofauna at deep-sea hydrothermal vents and cold seeps with particular reference to nematodes. Ph.D. thesis, National Oceanography Centre, Southampton

Gebruk A (2008) Holothurians (Holothuroidea, Echinodermata) of the northern Mid-Atlantic Ridge collected by the G.O. Sars MAR-ECO expedition with descriptions of four new species. Mar Biol Res 4:48–60

Gebruk AV, Krylova EM (2013) Megafauna of the Charlie–Gibbs Fracture Zone (northern Mid-Atlantic Ridge) based on video observations. J Mar Biol Assoc U K 93:1143–1150

Gebruk AV, Mironov AN (2006) Biogeografiya gidroterm Sredinno-Atlanticheskogo Khrebta. In: Vinogradov ME, Vereshchaka AL (eds) Ekosistemy atlanticheskikh gidroterm. Nauka, Moskva, pp 119–162

Gebruk AV, Chevaldonné P, Shank T, Lutz RA, Vrijenhoek RC (2000) Deep-sea hydrothermal vent communities of the Logatchev area (14°45′N, Mid-Atlantic Ridge): diverse biotopes and high biomass. J Mar Biol Assoc U K 80:383–393

Gebruk AV, Budaeva NE, King NJ (2010) Bathyal benthic fauna of the Mid-Atlantic Ridge between the Azores and the Reykjanes Ridge. J Mar Biol Assoc UK 90:1–14

Gollner S, Riemer B, Martínez Arbizu P, Le Bris N, Bright M (2010) Diversity of meiofauna from the 9°50′N East Pacific Rise across a gradient of hydrothermal fluid emissions. PLoS One 5:e12321

Gomes-Pereira JN, Auger V, Beisiegel K, Benjamin R, Bergmann M, Bowden D, Buhl-Mortensen P, De Leo FC, Dionísio G, Durden JM, Edwards L, Friedman A, Greinert J, Jacobsen-Stout N, Lerner S, Leslie M, Nattkemper TW, Sameoto JA, Schoening T, Schouten R, Seager J, Singh H, Soubigou O, Tojeira I, van den Beld I, Dias F, Tempera F, Santos RS (2016) Current and future trends in marine image annotation software. Prog Oceanogr 149:106–120

Goroslavskaya EI, Galkin SV (2011) Benthic fauna associated with mussel beds and shrimp swarms at hydrothermal fields on the Mid-Atlantic Ridge. Oceanology 51:69–79

Hestetun J, Fourt M, Vacelet J, Boury-Esnault N, Rapp H (2015) Cladorhizidae (Porifera, Demospongiae, Poecilosclerida) of the deep Atlantic collected during IFREMER cruises, with a biogeographic overview of the Atlantic species. J Mar Biol Assoc U K 95:1311–1342

Hilário A, Metaxas A, Gaudron SM, Howell KL, Mercier A, Mestre NC, Ross RE, Thurnherr AM, Young C (2015) Estimating dispersal distance in the deep sea: challenges and applications to marine reserves. Front Mar Sci 2:6

Horton T, Thurston MH, Duffy GA (2013) Community composition of scavenging amphipods at bathyal depths on the Mid-Atlantic Ridge. Deep Sea Res Part II Top Stud Oceanogr 98:352–359

Johnson DE (2019) Protecting the Lost City hydrothermal vent system: all is not lost, or is it? Mar Policy 103:103593

Jones DOB, Alt CHS, Priede IG, Reid WDK, Wigham BD, Billett DSM, Gebruk AV, Rogacheva A, Gooday AJ (2013) Deep-sea surface-dwelling enteropneusts from the Mid-Atlantic Ridge: their ecology, distribution and mode of life. Deep Sea Res II Top Stud Oceanogr 98:374–387

Kelley DS, Karson JA, Früh-Green GL, Yoerger DR, Shank TM, Butterfield DA, Hayes JM, Schrenk MO, Olson EJ, Proskurowski G, Jakuba M, Bradley A, Larson B, Ludwig K, Glickson D, Buckman K, Bradley AS, Brazelton WJ, Roe K, Elend MJ, Delacour A, Bernasconi SM, Lilley MD, Baross JA, Summons RE, Sylva SP (2005) A serpentinite-hosted ecosystem: the Lost City hydrothermal field. Science 307:1428–1433

Kelley DS, Früh-Green GL, Karson JA, Ludwig KA (2007) The Lost City hydrothermal field revisited. Oceanography 20:90–99

Klunder L, Stigter H, Lavaleye MSS, van Bleijswijk JDL, van der Veer HW, Reichart GJ, Duineveld GCA (2020) A molecular approach to explore the background benthic fauna around a hydrothermal vent and their larvae: implications for future mining of deep-sea SMS deposits. Front Mar Sci 7:134

Kongsrud JA, Budaeva N, Barnich R, Oug E, Bakken T (2013) Benthic polychaetes from the northern Mid-Atlantic Ridge between the Azores and the Reykjanes Ridge. Mar Biol Res 9:516–546

Larsen K, Błażewicz-Paszkowycz M, Cunha MR (2006) Tanaidacean (Crustacea: Peracarida) fauna from chemically reduced habitats – the Lucky Strike hydrothermal vent system, Mid-Atlantic Ridge. Zootaxa 1187:1–36

Linley TD, Alt CHS, Jones DOB, Priede IG (2013) Bathyal demersal fishes of the Charlie-Gibbs Fracture Zone region (49°–54°N) of the Mid-Atlantic Ridge: III. Results from remotely operated vehicle (ROV) video transects. Deep Sea Res Part II Top Stud Oceanogr 98:407–411

Marticorena J, Matabos M, Sarrazin J, Ramirez-Llodra E (2020) Contrasting reproductive biology of two hydrothermal gastropods from the Mid-Atlantic Ridge: implications for resilience of vent communities. Mar Biol 167:1–19

Martynov AV, Litvinova NM (2008) Deep-water Ophiuroidea of the northern Atlantic with descriptions of three new species and taxonomic remarks on certain genera and species. Mar Biol Res 4:76–111

Menegotto A, Rangel TF (2018) Mapping knowledge gaps in marine diversity reveals a latitudinal gradient of missing species richness. Nat Commun 9:4713

Molodtsova T, Sanamyan NP, Keller NB (2008) Anthozoa from the northern Mid-Atlantic Ridge and Charlie-Gibbs Fracture Zone. Mar Biol Res 4:112–130

Molodtsova TN, Galkin SV, Kobyliansky SG, Simakova UV, Vedenin AA, Dobretsova IG, Gebruk AV (2017) First data on benthic and fish communities from the Mid-Atlantic Ridge, 16°40′–17°14′N. Deep Sea Res Part II Top Stud Oceanogr 137:69–77. https://doi.org/10.1016/j.dsr2.2016.10

Morris KJ, Tyler PA, Murton B, Rogers AD (2012) Lower bathyal and abyssal distribution of coral in the axial volcanic ridge of the Mid-Atlantic Ridge at 45°N. Deep Sea Res Part I Oceanogr Res Pap 62:32–39

Myers AA, Cunha MR (2004) New and little known corophiidean amphipods from the 'Lucky Strike' hydrothermal vent, Mid-Atlantic Ridge. J Mar Biol Assoc U K 84:1019–1025

Niedzielski T, Høines A, Shields MA, Linley TD, Priede IG (2013) A multi-scale investigation into seafloor topography of the northern Mid-Atlantic Ridge based on geographic information system analysis. Deep Sea Res Part II Top Stud Oceanogr 98:231–243

Pabis K, Błażewicz-Paszkowycz M, Jóźwiak P, Barnes DKA (2015) Tanaidacea of the Amundsen and Scotia Seas: an unexplored diversity. Antarct Sci 27:19–30

Parin NV (1995) Deep-sea fishes, *Haptenchelys texis* (Synaphobranchidae) and *Pachycara thermophilum* (Zoarcidae) from the Broken Spur hydrothermal vent field (Mid-Atlantic Ridge). J Ichthyol 35:328–332

Perez JAA, dos Santos Alves E, Clark MR, Bergstad OA, Gebruk A, Azevedo Cardoso I, Rogacheva A (2012) Patterns of life on the southern Mid-Atlantic Ridge: compiling what is known and addressing future research. Oceanography 25:16–31

Priede IG, Bergstad OA, Miller PI, Vecchione M, Gebruk A, Falkenhaug T, Billett DSM, Craig J, Dale AC, Shields MA, Tilstone GH, Sutton TT, Gooday AJ, Inall ME, Jones DOB, Martinez-Vicente V, Menezes GM, Niedzielski T, Sigurðsson Þ, Rothe N, Rogacheva A, Alt CHS, Brand T, Abell R, Brierley AS, Cousins NC, Crockard D, Hoelzel AR, Høines Å, Letessier TB, Read

JF, Shimmield T, Cox MJ, Galbraith JK, Gordon JDM, Horton T, Neat F, Lorance P (2013) Does presence of a mid-ocean ridge enhance biomass and biodiversity? PLoS One 8:e61550

Radziejewska T, Błażewicz M, Włodarska-Kowalczuk M, Jóźwiak P, Pabis K, Węsławski JM (2022) Benthic biology in the Polish exploration contract area of the Mid-Atlantic Ridge: the knowns and the unknowns. A review. Front Mar Sci 9:898828

Riehl T, Lins Pereira L, Brandt A (2018) The effects of depth, distance, and the Mid-Atlantic Ridge on genetic differentiation of abyssal and hadal isopods (Macrostylidae). Deep Sea Res Part II Top Stud Oceanogr 148:74–90

Riou V, Duperron S, Dehairs HF, Bouillon S, Martin I, Colaço A, Santos RS (2010) Variation in physiological indicators in *Bathymodiolus azoricus* (Bivalvia: Mytilidae) at the Menez Gwen Mid-Atlantic Ridge deep-sea hydrothermal vent site within a year. Mar Environ Res 70:264–271

Rogacheva A, Gebruk A, Alt CHS (2013) Holothuroidea of the Charlie Gibbs Fracture Zone area, northern Mid-Atlantic Ridge. Mar Biol Res 9:587–623

Rybakova E, Galkin S (2014) Hydrothermal assemblages associated with different foundation species on the East Pacific Rise and Mid-Atlantic Ridge, with a special focus on mytilids. Mar Ecol 36:45–61

Sarrazin J, Legendre P, De Busserolles F, Fabri MC, Guilini K, Ivanenko VN, Morineaux M, Vanreusel A, Sarradin PM (2015) Biodiversity patterns, environmental drivers and indicator species on a high temperature hydrothermal edifice, Mid-Atlantic Ridge. Deep Sea Res Part II Top Stud Oceanogr 121:177–192

Shields MA, Blanco-Perez R (2013) Polychaete abundance, biomass and diversity patterns at the Mid-Atlantic Ridge, North Atlantic Ocean. Deep Sea Res Part II Top Stud Oceanogr 98:315–325

Shields MA, Glover AG, Wiklund H (2013) Polynoid polychaetes of the Mid-Atlantic Ridge and a new holothurian association. Mar Biol Res 9:547–553

Sigvaldadottir E, Desbruyères D (2003) Two new species of spionidae (Annelida: Polychaeta) from Mid-Atlantic Ridge hydrothermal vents. Cah Biol Mar 44:219–225

Spedicato A, Sánchez N, Pastor L, Menot L, Zeppilli D (2020) Meiofauna community in soft sediments at TAG and Snake Pit hydrothermal vent fields. Front Mar Sci 7:200

Sutton T, Porteiro F, Heino M, Byrkjedal I, Langhelle G, Anderson C, Horne J, Søiland H, Falkenhaug T, Godø O (2008) Vertical structure, biomass and topographic association of deep-pelagic fishes in relation to a mid-ocean ridge system. Deep Sea Res Part II Top Stud Oceanogr 55:161–184

Tabachnick K, Collins AG (2008) Glass sponges (Porifera, Hexactinellida) of the northern Mid-Atlantic Ridge. Mar Biol Res 4:25–47

Tchesunov AV (2015) Free-living nematode species (Nematoda) dwelling in hydrothermal sites of the north Mid-Atlantic Ridge. Helgol Mar Res 69:343–384

Teixeira S, Olu K, Decker C, Cunha RL, Fuchs S, Hourdez S, Serrão EA, Arnaud-Haond S (2013) High connectivity across the fragmented chemosynthetic ecosystems of the deep Atlantic Equatorial Belt: efficient dispersal mechanisms or questionable endemism? Mol Ecol 22:4663–4680

Thaler AD, Amon D (2020) 262 voyages beneath the sea: a global assessment of macro- and megafaunal biodiversity and research effort at deep-sea hydrothermal vents. PeerJ 7:e7397

Tyler PA, Paterson GJL, Sibuet M, Guille AF (1995) A new genus of ophiuroid (Echinodermata: Ophiuroidea) from hydrothermal mounds along the Mid-Atlantic Ridge. J Mar Biol Assoc U K 75:977–986

Vanreusel A, van den Bossche I, Thiermann F (1997) Free-living marine nematodes from hydrothermal sediments: similarities with communities from diverse reduced habitats. Mar Ecol Prog Ser 157:207–219

Vecchione M, Young RE, Piatkowski U (2010) Cephalopods of the northern Mid-Atlantic Ridge. Mar Biol Res 6:25–52

Washburn TW, Iguchi A, Yamaoka K, Nagao M, Onishi Y, Fukuhara T, Yamamoto Y, Suzuki A (2023) Impacts of the first deep-sea seafloor massive sulfide mining excavation tests on benthic communities. Mar Ecol Prog Ser 712:1–10

Yahagi T, Fukumori H, Warén A, Kano Y (2019) Population connectivity of hydrothermal-vent limpets along the northern Mid-Atlantic Ridge (Gastropoda: Neritimorpha: Phenacolepadidae). J Mar Biol Assoc UK 99:179–185

Zekely J, Van Dover CL, Nemeschkal HL, Bright M (2006) Hydrothermal vent meiobenthos associated with mytilid mussel aggregations from the Mid-Atlantic Ridge and the East Pacific Rise. Deep Sea Res Part I Oceanogr Res Pap 53:1363–1378

Zeppilli D, Vanreusel A, Pradillon F, Fuchs S, Mandon P, James T, Sarrazin J (2015) Rapid colonisation by nematodes on organic and inorganic substrata deployed at the deep-sea Lucky Strike hydrothermal vent field (Mid-Atlantic Ridge). Mar Biodivers 45:489–504

Zeppilli D, Leduc D, Fontanier C, Fontaneto D, Fuchs S, Gooday AJ, Goineau A, Ingels J, Ivanenko VN, Kristensen RM, Cardoso Neves R, Sanchez N, Sandulli R, Sarrazin J, Sørensen MV, Tasiemski A, Vanreusel A, Autret M, Bourdonnay L, Claireaux M, Coquillé V, De Wever L, Rachel D, Marchant J, Toomey L, Fernandes D (2018) Characteristics of meiofauna in extreme marine ecosystems: a review. Mar Biodivers 48:35–71

Seabed Communities

Maria Włodarska-Kowalczuk, Teresa Radziejewska, Katarzyna Jankowska, and Magdalena Błażewicz

1 Functioning: Energy Source and Trophic Pathways

Benthic systems on the Mid-Atlantic Ridge (MAR) are fuelled by two main energy sources: *in-situ* chemosynthetic production and/or photosynthetic production that is transported from the distant euphotic zone. Microbial chemolithoautotrophs (converting chemical energy and inorganic carbon to biomass) exploit energy sources provided by the hydrothermal vent fluids.

The off-axis Lost City Hydrothermal Field (LCHF) (cf. Chapter 3, and see below) is a unique system where serpentinization of ultramafic rocks (peridotites) produce warm (40–70 °C), alkaline (pH 9 and above) fluids rich in Ca, H_2 and CH_4 but poor in sulphides (Kelley et al. 2005; Von Damm 2001). Thermodynamic modelling pointed to metabolic reactions involving H_2 and CH_4 (particularly hydrogen oxidation, methanotrophy, sulphate reduction, and methanogenesis) as the predominant energy sources in ultramafic-hosted hydrothermal systems, and indicated that they can supply about twice as much chemical energy and potentially support a

M. Włodarska-Kowalczuk
Institute of Oceanology PAN, Sopot, Poland
e-mail: maria@iopan.gda.pl

T. Radziejewska (✉)
Institute of Marine and Environmental Sciences, University of Szczecin, Szczecin, Poland
e-mail: teresa.radziejewska@usz.edu.pl

K. Jankowska
Faculty of Environmental Engineering, Technical University of Gdańsk, Gdańsk, Poland
e-mail: kjank@pg.gda.pl

M. Błażewicz
Faculty of Biology, University of Łódź, Łódź, Poland
e-mail: magdalena.blazewicz@biol.uni.lodz.pl

T. Radziejewska et al. (eds.), *The Natural Environment of the Mid-Atlantic Ridge*, GeoPlanet: Earth and Planetary Sciences,
https://doi.org/10.1007/978-3-031-51865-2_6

much higher biomass than analogous systems hosted in basaltic rocks (McCollom 2007). The Lost City hydrothermal field appeared to be 'an energy bonanza in comparison with the majority of deep-sea habitats' (Lang and Brazelton 2020). Indeed, isotope analyses of organic carbon in active carbonate chimneys at Lost City evidenced microbial utilization of H_2 (in the form of sulphate reduction) and mantle inorganic carbon (Lang et al. 2012). The benthic macrofaunal invertebrates dwelling at the Lost City that most likely graze on microbial biofilms (Boetius 2005) are diverse, but not abundant (Kelley et al. 2005). However, high numbers of mussels, clams and gastropods recorded in 100-kyr old Lost City-like carbonates (Ghost City, 36°N) suggest that serpentinization-based hydrothermal systems are able to sustain high standing stocks of marine invertebrates (Lartaud et al. 2011). DeChaine et al. (2006) showed that few *Bathymodiolus* mussels present at the Lost City harbour endosymbionts closely phylogenetically related to those in mussels collected at basalt-based hydrothermal vents. Understanding the Lost City ecosystem functioning is of crucial importance as it can provide important insights into the origin of life on Earth and other planetary bodies in the solar system (Lang and Brazelton 2020).

At in-axis basalt-hosted hydrothermal fields, seawater-basalt interactions produce fluids that are hot (~350 °C), acid (pH 3–5) and enriched in H_2S, H_2, CH_4, Mn and other metals (Van Dover 2000). Energy sources for primary production are oxidative reactions (predominantly sulphide oxidation) at lower temperatures (below ~38 °C) and reductive reactions within the walls of vent chimneys (McCollom and Shock 1997). Extreme environmental conditions constrain the invertebrate species pools and favour simple food webs with short food chains (Colaço et al. 2002). The basic trophic structure is similar across different fields on the MAR (including Broken Spur located in the Polish contract area), with chemosynthetic producers at the base of the food web and two main types of primary consumers: endosymbiont-hosting (e.g., *Bathymodiolus* mussels) or feeding on free-living bacteria (e.g., *Rimicaris* shrimps). Predators depend on one (the crab *Sergonzacia* preferring shrimps, the gastropod *Phymporhychus* feeding almost exclusively on mussels) or both groups of primary consumers. Deep-sea bathyal species making incursions into the vent fields act as top predators (Colaço et al. 2002). These pathways of energy flow have been observed in a number of studies performed at different sites (including the Broken Spur) and employing *in-situ* and experimental observations as well as analyses of chemical biomarkers (stable isotopes and fatty acid profiles) (e.g., Gebruk et al. 2000; Vereshchaka et al. 2000; Colaço et al. 2007). Portail et al. (2018) compared 13 faunal assemblages in the Azores Triple Junction and demonstrated that benthic consumers were highly flexible (in terms of their dependence on the composition of producers and metabolic diversity), which secured the efficient functioning and stability of trophic systems across different MAR hydrothermal vent localities.

Organic matter produced at hydrothermal fields can be advected at least 2 km away from the site of active venting (Roth and Dymond, 1989) and, to some extent, support benthic communities therein (Erickson et al. 2009). However, functioning of non-vent deep-sea communities depends essentially on the organic matter transported from the sea surface where light penetration allows photosynthetic production. Reid et al. (2012) indicated two main trophic pathways linking pelagic producers and benthic consumers in non-hydrothermal MAR benthic habitats: (1) invertebrate consumers (detritus and suspension feeders) dependent on the downward flux of phytodetritus to the seafloor; (2) mobile predator and scavenging fishes and crustaceans exploiting aggregations of meso- and bathypelagic biomass associated with topographic features of the ridge (Sutton et al. 2008).

2 Benthic Abundance and Biomass

Since their discovery in the late 1970s, hydrothermal vent ecosystems have often been referred to, both in scientific publications and in popular-science literature, as 'oases of life' on the generally depauperate deep-sea floor (e.g., Carney 1994; wwf.panda.org/our_work/our_focus/oceans_practice/deep_sea/vents_seeps/). In fact, the richness of life associated with the newly discovered hydrothermal vent fields was the most striking feature recorded and reported by the researchers who discovered the vents on the East Pacific Rise (www.whoi.edu/oceanus/feature/the-discovery-of-hydrothermal-vents/). The oasis analogy is justified also with respect to hydrothermal fields explored on the Mid-Atlantic Ridge (MAR), primarily in view of the mass occurrence of megafaunal organisms, visible in numerous images collected during exploration of different fields and sites within them (Galkin 2006 and references therein; Van Dover 2014).

A summary of benthic abundance and standing stock data collected from the MAR up to 2005 (Galkin 2006) focuses primarily on the dominant and foundation species (Rybakova and Galkin 2015). These include bresilioid (Bresilioidea) shrimps found in all the MAR fields explored. They are most prominently represented by the landscape-forming species *Rimicaris exoculata* (family Alvinocarididae) with abundances of up to 2500–3000 ind./m^2 and biomass of up to 1–1.5 kg/m^2; such abundances have been reported from, *inter alia*, the Broken Spur field in the Polish contract area. The *Rimicaris* occur frequently in associations with other shrimps (*Chorocaris chacei*, up to 50 ind./m^2 in the TAG field; Gebruk et al. 1997), *Segonzacia mesatlantica* crabs, and ectinosomatid harpacticoids; Segonzac (1992) estimated the biomass of those associations at up to 1.2 kg/m^2.

Another group of foundation and dominant species is formed by bivalves of the families Mytilidae (genus *Bathymodiolus*) and Vesicomyidae (genus *Calyptogena*) (Galkin 2006). In terms of abundance and biomass, the two *Bathymodiolus* species: *B. azoricus* and *B. puteoserpentis* known from the MAR, are very unevenly distributed among individual fields. While *B. azoricus* is more abundant at shallower fields (Rainbow, Lucky Strike, Menez Gwen, with abundances of up to 10^3 ind./m^2), *B. puteoserpentis* forms very dense and extensive beds in other (Logatchev, Snake Pit and Broken Spur), with more than 70 kg/m^2 biomass at the Logatchev field, but is much less abundant and occur in local patches only at Broken Spur. The vesicomyid bivalves are primarily represented by *Calyptogena* aff. *kaikoi*, found in aggregations of a spatial extension, e.g., at the Logatchev field, of up to 100 m^2 (Galkin 2006).

Important, in terms of abundance, are gastropods: for instance, limpets *Peltospira smaragdina* were reported with abundances of up to several hundred individuals per m^2, while gastropods of the genus *Phymorchynchus* (e.g., *Ph. moskalevi*), found, for instance, in the inactive part of the TAG field, occurred at abundances of up to 3–10 ind./m^2 (Galkin 2006).

As stated by Van Dover (2014), the hydrothermal vent ecosystems are dominated by the so-called holobiont taxa, i.e., organisms that host symbiotic chemoautotrophic microorganisms. They often occur as the foundation species forming complex 3-dimensional habitats (Gerdes et al. 2019), for example those created by the mytilid and vesicomyid bivalves mentioned above as well as the chaetopterid polychaetes (genus *Chaetopterus*) occurring in masses away from active smokers. Those aggregations serve as a substratum for the growth of epizoic microorganisms, refugia for juvenile invertebrates, and a habitat for associated organisms, including the meiobenthos (nematodes, harpacticoids, halacarids)—occasionally, e.g., in Broken Spur, found in abundance, primary consumers (e.g., limpets such as *Peltospira smaragdina* and *Lepetodrilus atlanticus* grazing on microbial biofilms) as well as secondary and tertiary consumers such as scavenging and predatory crustaceans and fishes. The *Bathymodiolus* bed-associated macrofaunal abundance at Logatchev has been reported by Van Dover and Doerries (2005) as ranging (by sample volume, collected with 'benthic pots') within 209–2390 ind./dm^3.

Holobiont taxa typically dominate the biomass of habitable diffuse flow regions (in the vicinity of active smokers) and show abrupt transitions from one dominant species to another, reflecting different interspecific tolerances for thermal and chemical conditions and to biotic interactions (facilitation, competition, predation) (Van Dover 2014). The total biomass values, indicated above (aggregations of *Rimicaris exoculata* or *Bathymodiolus puteoserpentis*) are typically very high at vents. Beyond the periphery of a vent field, the faunal abundance and biomass values are usually much lower, the megafauna being represented by solitary large anemones (e.g., *Maractis rimicarivora*, which occurs also on hard substrata in the vicinity of venting; Fabri et al. 2011) and gorgonian corals (Van Dover 2014).

Observations on the macrofauna at the Lost City (Shank 2005 at oceanexplorer.noaa.gov/explorations/05lostcity/background/macrofauna/macrofauna.html) showed a high biodiversity ('over 70 potential species'). The biomass (wet weight) was apparently lower compared to a typical hydrothermal vent on the MAR. The major contributors to the biomass at Lost City are the larger more mobile megafauna, including the (grouper-like) wreckfish, cut-throat eels, and large red geryonid crabs, all readily visible around the spires. Shank (op. cit.) reported also on a sharp difference between the fauna of vent and non-vent habitats at the Lost City, whereby typical vent areas are sparsely populated, whereas non-venting habitats less than a few metres away (e.g., on the sides of inactive solidified carbonate structures, sedimented areas, and breccia cap rock just to the north of the field) are dominated by hard corals (*Lophelia pertusa* and *Desmophyllum*), octocorals (gorgonians), galatheid crabs, turrid gastropods, foraminifera, pteropods, urchins, asteroids, ophiuroids, and typical deep-sea barnacles.

The soft sediment-covered areas away from active venting have been reported as supporting fairly abundant (considering the low general abundances of deep-sea infauna; cf. Radziejewska 2014) meiobenthos (e.g., 11–35 and 72–86 ind./10 cm^2 at TAG and Snake Pit, respectively). Faunal density responses appear to be a function of stress level. For example, at Middle Valley (2410 m, the Juan de Fuca hydrothermal vent field in the Pacific), extremely hot sediments (e.g., 94 °C at 5 cm into the sediment column) support very few macrofauna, whereas moderately warm sediments inhabited by vesicomyid bivalves may support elevated macrofaunal densities (16,500 ind./ m^2) relative to those in microbial mats (6840 ind./m^2), hot sediment (1690 ind./m^2), and control (not heated) sediments (2218 ind./m^2) (Bernardino et al. 2012). However, no comparative data are available from the MAR as yet.

3 Microbial Communities: Lost City

The Lost City hydrothermal field on the Mid-Atlantic Ridge (Fig. 1) supports dense microbial life on the lofty calcium carbonate chimney structures (McGonigle et al. 2020). The dense microbial biofilms of Lost City chimneys are fuelled by the carbon and energy released by serpentinization of the underlying ultramafic rock (Schrenk et al. 2004; Brazelton et al. 2006, 2010, 2011; Lang et al. 2018).

Fig. 1 An assemblage of carbonaceous 'chimneys' at the Lost City hydrothermal field (750-900 m depth), as photographed from the ROV Victor 6000. Source: IFREMER, https://image.ifremer.fr/data/00569/68094/

Prior to the discovery of chemoautotrophic ecosystems in seafloor hydrothermal systems in the 1970–1980s, it had been accepted that all ecosystems ultimately depend on energy from the sunlight. The discovery implied a possibility for life to exist outside Earth to include potential habitats that are magmatically active even if they lack abundant sunlight (Lang and Brazelton 2020).

In addition, the high concentrations of H_2 and appropriate catalysts in a hydrothermal system can lead to the abiotic synthesis of organic molecules which provide a source of food for life and could have also played a major role in early biochemical evolution (Mccollom and Seewald 2007; Martin et al. 2008). Serpentinites were probably more abundant and active on the early Earth than they are now (Sleep et al. 2004), and serpentinization has probably occurred to some degree on all rocky planetary bodies in the solar system (Sleep et al. 2004). These systems are thought to be present on icy worlds, such as Jupiter's moon Europa and Saturn's moon Enceladus (Kargel et al. 2000; Oze et al. 2015; Glein et al. 2015; Waite et al. 2017). Therefore, a better understanding of the biological opportunities and challenges in serpentinizing systems provides important insights into the potential for the origin and persistence of life on Earth and elsewhere in the solar system (Lang and Brazelton 2020).

Research of the microbial biofilm communities that live in the chimneys of the Lost City hydrothermal field can help to decide whether serpentinization and its associated geochemical reactions are sufficient to support the origin and evolution of life independently of other geological processes, or if the products of serpentinization must be mixed with other materials to meet all of the requirements for continuous habitability (Stüeken et al. 2013).

H_2 is highly abundant (up to 14 μM) in the Lost City chimneys, which contributes to high concentrations of methane (CH_4) and formate (Kelley et al. 2005; Proskurowski et al. 2006, 2013; Lang et al. 2010). Consequently, the Lost City biofilm communities are dominated by organisms likely to consume H_2, CH_4 and formate (Brazelton et al. 2006, 2011; Lang et al. 2018).

However, the Lost City chimneys are mixing zones where anoxic, high pH subsurface fluids mix with oxic seawater. The remarkable density of microbes in the chimneys [up to 10^9 cells per gram (Schrenk et al. 2004)] is a product of the many physical and chemical gradients formed within the porous matrix of the chimneys when these two very different kinds of water are mixed together. Wherever and whenever these gradients can be established, the potential for biochemical and microbial activity is high (Summit and Baross 2001; McCollom and Seewald 2007).

Interestingly, as shown by the Integrated Ocean Drilling Project (IODP) 304 and 305 (Lang and Brazelton 2020), data from the first microbiological investigation of rock cores collected of the Atlantis Massif, a large seafloor mountain located near the Lost City chimneys, have revealed a much lower biological content compared with the Lost City chimneys. Microbial cells are nearly undetectable by microscopy in many serpentinite core samples and are typically present at densities approximately 100 cells/cm^3 (Früh-Green et al. 2018). Similarly, low cell densities (below the detection limit of less than 10^3 cells/cm^3) were also found in gabbroic core samples drilled on an earlier expedition (Mason et al. 2010). These low microbial densities suggest that the habitability of the serpentinite subsurface is limited by physical and/or geochemical conditions. However, it remains possible that the active microbial populations in serpentinite subsurface habitats are present in sparse but dense patches, most likely concentrated along pathways of the fluid flow through the basement rocks (Lang and Brazelton 2020).

During the last expedition in 2018 (https://lostcity.biology.utah.edu) (the public data available under BCO-DMO Project No. 658604), an attempt has been made to investigate in a comprehensive way the relationship between factors relevant for understanding and predicting the habitability of serpentinite-hosted environments. The research has been structured by the requirements for habitability as summarized by Cockell et al. (2016) such as a solvent, physico-chemical conditions, energy, major elements needed for life (C, H, N, O, P, S), and additional required elements, such as trace metals (Lang and Brazelton 2020). Lang and Brazelton (2020) concluded, *inter alia*, that the evolution and maintenance of a robust biological ecosystem requires continuous habitability over geological time scales (Cockell et al. 2016). The absence of certain elemental requirements (e.g., a solvent, energy, appropriate physico-chemical conditions, major and minor elements for life) will preclude life from being able to maintain its complex state. But simply meeting the criteria may also be insufficient to achieve habitability. Rather, amounts of resources also matter for each parameter individually and in combination (Hoehler 2007). In serpentinizing systems, life may not be limited by a single limiting factor, but instead by combinations of factors that cannot be overcome simultaneously. These observations complement well the studies on microbial biofilm communities that live in the chimneys of the Lost City hydrothermal field presented by McGonigle et al. (2020).

Previous studies (Schrenk et al. 2004) indicated that microbes in interiors of chimneys where the temperatures of venting fluids can reach 95 °C, and the pH of the fluids can be as high as 11 (Butter et al. 2001), are dominated by a single archaeal phylotype, the Lost City Methanosarcinales (Schrenk et al. 2004; Kelley et al. 2005; Brazelton et al. 2006, 2010, 2011; Lang et al. 2018). In contrast, the chimney exteriors host a more complex microbial community, including organisms involved in the oxidation of sulphur and methane (e.g., *Methylomonas*, *Thiomicrospira*). It was noted that these organisms likely thrive in the mixing zones, where they can take advantage of the cooling effect of the seawater and more efficient electron acceptors (e.g., oxygen), but still access the products of serpentinization supplied by venting fluids.

Previous studies (Lang et al. 2012, 2018) showed, too, that much of the microbial biomass at the Lost City is derived from carbon that originated deep in the Earth's subsurface. In most ecosystems, inorganic carbon (CO_2) serves as the starting carbon source for primary production. However, the Lost City fluids contain extremely low concentrations of dissolved inorganic carbon (DIC) due to its reduction to hydrocarbons and its rapid precipitation as calcium carbonate at a pH above ~9 (Kelley et al. 2005; Proskurowski et al. 2013). The organic acid formate has been proposed to be an alternative primary carbon source; it is present in high concentrations (36 to 158 μM) in Lost City fluids and is expected to form abiotically in serpentinizing fluids (Lang et al. 2010; McDermott et al. 2015; Lang et al. 2018).

Formate is unable to enter carbon fixation pathways directly and needs to be converted to CO_2 for autotrophic metabolism (Yishai et al. 2016). The enzyme formate dehydrogenase catalyzes the reversible oxidation of formate to CO_2.

The Methanosarcinales phylotype has previously been shown to dominate the anoxic, interior zones of Lost City chimneys (Schrenk et al. 2004; Brazelton et al. 2010, 2011), yet it appears to be unable to use one of the most abundant carbon sources, formate. Although novel, previously undiscovered formate metabolism genes may exist, no archaeal formate dehydrogenase gene sequences (fdhA, fdhB) or fdhC sequences affiliated with methanogens were detected in the metagenome (McGonigle et al. 2020). These studies also identify the genomic potential of the *Chloroflexi* and *Sulfurovum* populations to utilize formate, which may be required to make mantle-derived carbon available to the rest of the chimney ecosystem, although the *Chloroflexi* is seven times less abundant than the Methanosarcinales in the chimney sample. However, it was noted that the potential explanation is that *Chloroflexi* species are highly active and able to rapidly cycle carbon while maintaining a low abundance in the biofilm community. Alternatively, the *Chloroflexi* population may be more abundant in subsurface habitats underlying the chimneys, where formate is expected to be generated (Lang et al. 2018).

In conclusion, it should be noted how valuable the Lost City hydrothermal field on the Mid-Atlantic Ridge is. Getting to know this environment will help to understand even better how multiple extreme limitations of life interact to restrict, or perhaps promote, habitability in unusual and complex environments.

4 Endemism on the Mid-Atlantic Ridge (MAR)

Endemism is understood as a restricted geographical distribution of a species to a well-defined area. The fauna associated exclusively with hydrothermal vents and absent outside the vent fields is classified as 'vent-obligate'. The vent-obligate fauna components, however, are often widely distributed, although their geographical ranges are usually disjunct (Mironov et al. 2002; Gollner et al. 2015; Baco et al. 2016). The vent-obligate fauna is accompanied by 'peripheral' or 'penetrating' fauna, represented by taxa non-obligatorily associated with vent fields. In contrast to the vent-obligate fauna, it is often recorded outside hydrothermal fields (Desbruyères et al. 2000; Galkin and Sagalevich 2016).

4.1 Fauna of Atlantic and Pacific Hydrothermal Sites

In the Pacific and the Atlantic, the vent-obligate fauna is represented by distinct sets of taxa; this is illustrated in Fig. 2 showing a clear separation between clusters of the Pacific and Atlantic vent faunas. Of the 56 vent-obligate families, 28 (50%) are

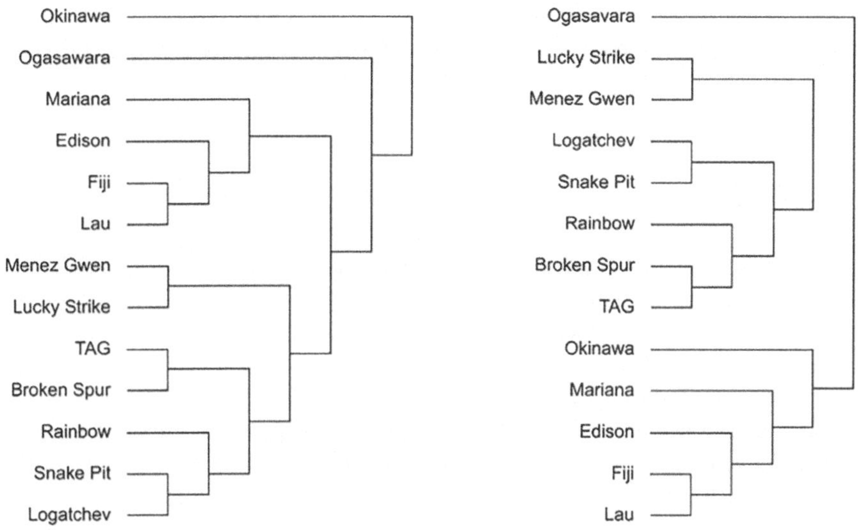

Fig. 2 Dendrograms of the Bray-Curtis similarity (presence-absence transformation, group average sorting) for vent-obligate fauna of Pacific and Atlantic sites based on genera and family level listed by Mironov et al. (2002); family level (left) and genus level (right)

present exclusively in the Pacific, 12 (21.4%) exclusively in the Atlantic, and 16 (28.6%) are common for both oceans. A relatively similar distinctness of the Pacific from the Atlantic fauna is observed on the genus level. Sixty of the 98 (61.2%) vent-obligate genera occur only in the Pacific, 26 (26.5%) in the Atlantic, and only 12 (12.2%) inhabit both oceans, including *Phymorhynchus* (Conidae), *Laeviphitus* (Elachisinidae), *Pseudorimula* (Lepetodrilidae), *Shinkailepas* (Phenacolepadidae), *Xylodiscula* (Xylodisculidae), *Bathymodiolus* (Mytilidae), *Copidognathus* (Halacaridea), *Alvinocaris* and *Chorocaris* (Alvinocarididae).

4.2 Vent-Obligate and Peripheral Hydrothermal Vent Fauna of the Mid-Atlantic Ridge (MAR)

The undersampling of the vent-peripheral fauna on the Mid-Atlantic Ridge (MAR) hampers its reliable comparison against the vent-obligate fauna, although distinct differences do exist (vent-obligate taxa are not found away from the vent field). On the other hand, earlier observations suggest that the MAR peripheral fauna is not markedly different from the fauna inhabiting other deep-sea habitats (Larsen et al.

2006). Those observation, however, are based on taxonomic surveys, while with an appropriately selected molecular approach, some subtle differences may be observed in the peripheral vent fauna along the MAR (Silva 2016).

4.3 Endemicity of the MAR Vent-Obligate Fauna

The vent-obligate fauna on the MAR shows a moderate degree of endemicity only. Most members of the fauna are widely distributed along the MAR on account of their good swimming ability, the presence of planktonic larvae or due to passive migration aided by larger, actively migrating vent species (Jollivet 1996; Teixeira et al. 2011; Hilário et al. 2015).

At the present state of knowledge, only five families are considered unique for individual vent fields: the sea anemones Actinostolidae (found at TAG), the gastropods Neolepetopsidae (Lucky Strike), the monoplacophorans Neopilinidae (Menez Gwen), and two amphipod families—the Amphilochidae (Lucky Strike) and the Stegocephalidae (Snake Pit).

Sixty-one species present in seven location of the MAR: Menez Gwen, Lucky Strike, Rainbow, Broken Spur, TAG, Snake Pit and Logatchev (Mironov et al. 2002) are considered vent-obligate fauna, and half of them (30 species or 49%) are endemic (Fig. 3, Tables 1 and 2). Furthermore, nine species (14.7%) were found at two sites, nine others (14.7%) at three sites, whereas six species (9.8%) were recorded at four stations and five species (8.2%) at three sites.

Generally, a high level of endemism is observed in crustacean amphipods and alvinocarids shrimps, with four species (80%) recorded at only one site each.

The benthopelagic *Rimicaris exoculata* (Alvinocaridae), present in swarms at MAR hydrothermal vents and the benthic bivalves *Bathymodiolus azoricus* and *B. puteoserpentis* (Mytilidae) are not endemic to any single vent field on the MAR, but because they have not been recorded in non-Atlantic vent fields, they are recognized as the iconic megafauna of the MAR hydrothermal vents. Where their ranges overlap, like in the Broken Spur vent field (in the Polish contract area), they were observed to hybridise (O'Mullan et al. 2001).

Incidentally, a new shrimp species, *Keldyshicaris vavilovi* was discovered in 2001 in the Broken Spur vent field (Lunina and Vereshchaka 2010). Since it has not been so far reported from any other location, it has to be regarded as strictly endemic.

Endemic taxa are also known within the macro- and meiofauna. For example, in the latter, the siphonostomatoid copepod family Dirivultidae is known exclusively from vent habitats (e.g., Gollner et al. 2015). Four known vent species of the Halacaroidea (Acarina) and nine of the 13 (69.2%) species of Dirivultidae (Copepoda) are known from only one location.

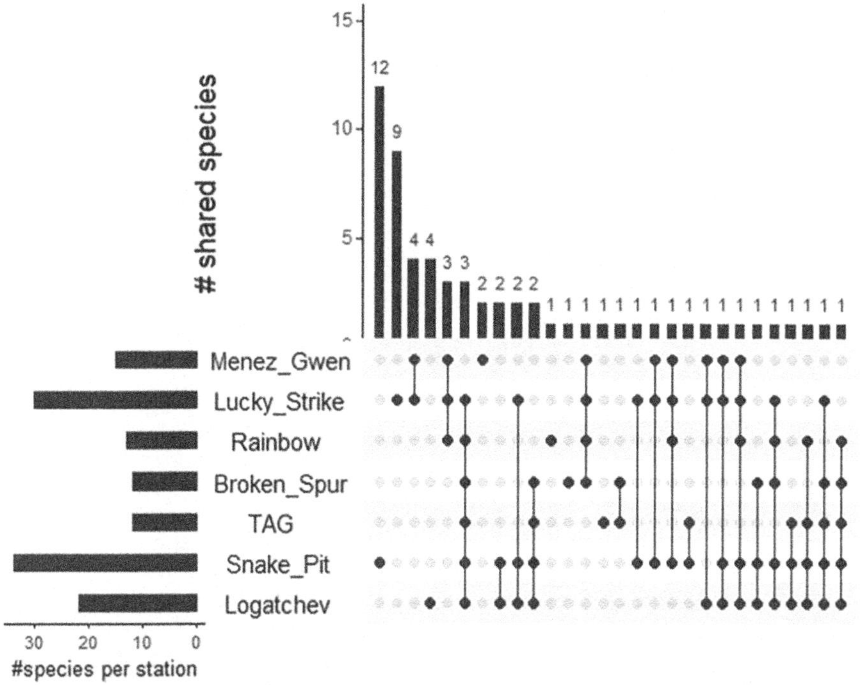

Fig. 3 UpSet matrix design bi-plots representing obligatory vent species recorded on MAR (upper plot) across seven vent sites. Points connected by a line indicate the shared species between the different sites, for which the total number of species is reported on top of the bar. Single dots represent unique species per site

Table 1 Vent-obligate fauna recorded in seven locations of MAR (n, number of species)

	Species	Endemic species (30)	
MAR vent field	n	n	%
Snake Pit	34	12	19.7
Lucky Strike	30	9	17.7
Logatchev	22	4	6.5
Menez Gwen	15	2	3.2
Rainbow	13	1	1.6
Broken Spur	12	1	1.6
TAG	12	1	1.6

Table 2 Taxa characteristic of MAR vent fields studied

Broken Spur	Lucky Strike	Snake Pit	Logatchev	Menez Gwen	Rainbow	TAG
Alvinocaris markensis	*Alvania stenolopha*	*Alvinocaris markensis*	*Alvinocaris* aff. *muricula*	*Alvania stenolopha*	*Alvinocaris* sp. 2	*Alvinocaris markensis*
Amathys lutzi	*Alvinocaris aff stactophila*	*Amathys lutzi*	*Alvinocaris markensis*	*Anisopes heteroscela*	*Amathys lutzi*	*Archinome aff rosacea*
Bathymodiolus puteoseprentis	*Alvinocaris markensis*	*Andaniotes ingens*	*Alvinocaris* sp. 1	*Aphotopontius atlanteus*	*Bathymodiolus azoricus*	*Bathymodiolus puteoserpentis*
Chorocaris chacei	*Amathys lutzi*	*Aphotopontius forcipatus*	*Amathys lutzi*	*Bathymodiolus azoricus*	*Chorocaris chacei*	*Chorocaris chacei*
Mirocaris fortunata	*Anisopes heteroscela*	*Archinome aff. rosacea*	*Anisopes heteroscela*	*Halacarellus prolongus*	*Laeviphitus desbruyeresi*	*Maractis rimicarivora*
Mirocaris keldyshi	*Aphotopontius atlanteus*	*Bathymodiolus puteoseprentis*	*Archinome* aff. *rosacea*	*Laeviphitus desbruyeresi*	*Leptodrilus atlanticus*	*Mirocaris keldyshi*
Ophictenella acies	*Aphotopontius temperatus*	*Candelabrum srepentari*	*Bathymodiolus puteoseprentis*	*Leptodrilus atlanticus*	*Mirocaris fortunata*	*Ophictenella acies*
Rimicaris exoculata	*Bathymodiolus azoricus*	*Chorocaris chacei*	*Calyptogena* aff. *kaikoi*	*Luckia striki*	*Ophictenella acies*	*Phymorhynchus moscalevi*
Rimipontius mediospinifer	*Bouvierella curtirama*	*Halacarellus auzendei*	*Candelabrum srepentari*	*Lurifax vitreus*	*Phymorhynchus moscalevi*	*Rimicaris exoculata*
Sengonzacia mesatlantica	*Chorocaris chacei*	*Leptodrilus atlanticus*	*Chorocaris chacei*	*Mirocaris fortunata*	*Rimicaris exoculata*	*Sengonzacia mesatlantica*
Stygiopontilus cladarus	*Dendronotus comteti*	*Lirapex* sp. A	*Copidognathus nautilei*	*Peltospira smaragdina*	*Sengonzacia mesatlantica*	*Stygiopontilus pectinatus*

Stygiopontilus pectinatus	*Gigantopsis alvina*	*Mirocaris keldyshi*	*Mirocaris keldyshi*	*Protolira valvatoides*	*Shinkailepas briandi*	
	Halacarellus alvinus	*Ophictenella acies*	*Ophictenella acies*	*Rocopella segonzaci*	*Xylodiscula analoga*	
	Laeviphitus desbruyeresi	*Peltospira smaragdina*	*Peltospira smaragdina*	*Shinkailepas briandi*		
	Leptodrilus atlanticus	*Phymorhynchus carinatus*	*Phymorhynchus carinatus*	*Xylodiscula analoga*		
	Lirapex costellata	*Phymorhynchus moscalevi*	*Phymorhynchus moscalevi*			
	Luckia striki	*Phymorhynchus ovatus*	*Phymorhynchus ovatus*			
	Lurifax vitreus	*Protolira valvatoides*	*Pseudorimula midatlantica*			
	Mirocaris fortunata	*Pseudorimula midatlantica*	*Rimicaris exoculata*			
	Ophictenella acies	*Rimicaris exoculata*	*Rimipontius mediospinifer*			
	Paralepetopsis ferrugivora	*Rimipontius mediospinifer*	*Sengonzacia mesatlantica*			
	Peltospira smaragdina	*Sengonzacia mesatlantica*	*Shinkailepas briandi*			
	Phymorhynchus ovatus	*Shinkailepas briandi*				
	Protolira valvatoides	*Steleuthera ecoprophycea*				

(continued)

Table 2 (continued)

Broken Spur	Lucky Strike	Snake Pit	Logatchev	Menez Gwen	Rainbow	TAG
	Pseudorimula midatlantica	*Stygiopontius bulbisetiger*				
	Rimicaris exoculata	*Stygiopontius cladarus*				
	Shinkailepas briandi	*Stygiopontius latulus*				
	Stygopontilus rimivagus	*Stygiopontius mirus*				
	Sutilizona pterodon	*Stygiopontius pectinatus*				
	Xylodiscula analoga	*Stygiopontius regius*				
		Stygiopontius serratus				
		Stygiopontius teres				
		Sutilizona pterodon				

5 Natural Variability

The greatest challenge in any environmental assessment is to disentangle effects of anthropogenic pressures from those resulting from natural variability (e.g., Osenberg et al. 1994). This challenge becomes more acute in areas which are greatly undersampled, particularly along the temporal axis. This is the case of the Mid-Atlantic Ridge (MAR).

Natural variability in deep-sea communities is shaped by a number of abiotic and biotic factors operating at various spatial and temporal scales (Glover et al. 2010). The natural processes at hydrothermal vents can lead to disturbances operating with different frequencies, spatial range and severity ranging from rather negligible effects of regular tidal fluctuations in fluid flows to unpredictable and catastrophic regimes resulting from tectonic activity and volcanic eruptions and leading to local extinctions (Van Dover 2014). The range of biotic factors bringing about natural variability in communities includes the growth rates of organisms, their life cycle (reproduction, maturation), and dispersal capabilities. Typically, these are biological traits that underpin temporal persistence of the communities in the face of natural disturbance (Van Dover 2014). The structure of the natural processes that shape the chemoautotrophic production (the basis of the vent ecosystems functioning) spanning a range of increasing spatial and temporal scales has been summarized by Le Bris et al. (2019).

The large-scale spatial variability in biological communities of the MAR has been treated fairly extensively, based on the available data: for example, Galkin (2006) described differences and similarities in benthic communities between several active hydrothermal vents (including the Broken Spur in the Polish contract area) and reported on some differences in the set of dominants and associated species. The natural variability on the scale of a single vent field was tackled by several studies. For example, spatial variability in vent communities of the Menez Gwen and Lucky Strike fields was treated by Desbruyères et al. (2001). They ascribed the variations observed in the structure and composition of the communities along the depth gradient to changes in vent fluid toxicity (metallic and sulphide content) and suspended mineral particles. Van Dover (2014) treats this spatial variability as zonation, with the holobiont (symbiont-supporting) taxa dominant in the habitable diffuse flow regions and abrupt transitions observed with a distance from the vent. While the total biomass of benthic organisms is very high at vents, at a hydrothermal periphery the biomass and abundances are greatly reduced. Nevertheless, information on patterns of natural spatial variability in many other MAR fields, including those in the Polish contract area, except for the information provided by Galkin (2006) is missing.

Knowledge of the temporal dimension in natural variability of deep-sea systems, particularly those of the MAR, is rather fragmentary. Few studies that included repeated visits and/or observations at vent sites differed in temporal range and resolution and focused mostly on invertebrates visible on underwater images (large epifauna). For example at Lucky Strike vent field, day-to-day variations in

Bathymodiolus azoricus mussel assemblages were monitored for a period of 48 days (Sarrazin et al. 2014), while another study focused on crustacean behaviour monitored for a week each month over a total period of 9 months (Matabos et al. 2015). A long term monitoring station was established in the Lucky Strike vent field in 2010 (EMSO-Azores multidisciplinary seafloor observatory, http://www.emso-fr.org/EMSO-Azores). Glover et al. (2010) reviewed the current literature on temporal trends in different habitats and regions in the deep-sea and stated that slow-spreading venting sites (as MAR) and associated biological systems are relatively stable on decadal to century timescales. Establishing a long-term monitoring programme in the Polish contract area is needed to provide essential ecological baseline information necessary to recognise and predict ecosystem responses to anthropogenic impacts in the area.

References

Baco AR, Etter RJ, Ribeiro PA, von der Heyden S, Beerli P, Kinlan BP (2016) A synthesis of genetic connectivity in deep-sea fauna and implications for marine reserve design. Mol Ecol 25:3276–3298. https://doi.org/10.1111/mec.13689

Bernardino AF, Levin LA, Thurber AR, Smith CR (2012) Comparative composition, diversity and trophic ecology of sediment macrofauna at vents, seeps and organic falls. PLoS One 7:e33515

Boetius A (2005) Lost City life. Science 307(5714):1420–1422

Brazelton WJ, Schrenk MO, Kelley DS, Baross JA (2006) Methane- and sulfur-metabolizing microbial communities dominate the Lost City hydrothermal field ecosystem. Appl Environ Microbiol 72:6257–6270. https://doi.org/10.1128/AEM.00574-06

Brazelton WJ, Ludwig KA, Sogin ML, Andreishcheva EN, Kelley DS, Shen C, Edwards RL, Baross JA (2010) Archaea and bacteria with surprising microdiversity show shifts in dominance over 1000-year time scales in hydrothermal chimneys. Proc Natl Acad Sci 107:1612–1617. https://doi.org/10.1073/pnas.0905369107

Brazelton WJ, Mehta MP, Kelley DS, Baross JA (2011) Physiological differentiation within a single-species biofilm fueled by serpentinization. mBio 2:e00127-11. https://doi.org/10.1128/mBio.00127-11

Butter DA, Lilley MD, Kelley DS, Karson JA, Blackman DK, Früh GL, Olson EJ, Schrenk MO, Roe KK, Lebon GT, Rivizzigno P, Party A-S (2001) An off-axis hydrothermal vent feld near the Mid-Atlantic Ridge at 30°N. Nature 412:145–149

Carney RS (1994) Consideration of the oasis analogy for chemosynthetic communities at Gulf of Mexico hydrocarbon vents. Geo-Mar Lett 14:149–159

Cockell CS, Bush T, Bryce C, Direito S, Fox-Powell M, Harrison JP, Lammer H, Landenmark H, Martin-Torres J, Nicholson N, Noack L, O'Malley-James J, Payler SJ, Rushby A, Samuels T, Schwendner P, Wadsworth J, Zorzano MP (2016) Habitability: a review. Astrobiology 16:1–29. https://doi.org/10.1089/ast.2015.1295

Colaço A, Dehairs F, Desbruyères D (2002) Nutritional relations of deep-sea hydrothermal fields at the Mid-Atlantic Ridge: a stable isotope approach. Deep Sea Res Part I Oceanogr Res Pap 49:395–412

Colaço A, Desbruyères D, Guezennec J (2007) Polar lipid fatty acids as indicators of trophic associations in a deep-sea vent system community. Mar Ecol 28:15–24

DeChaine EG, Bates AE, Shank TM, Cavanaugh CM (2006) Off-axis symbiosis found: characterization and biogeography of bacterial symbionts of *Bathymodiolus* mussels from Lost City hydrothermal vents. Environ Microbiol 8:1902–1912

Desbruyères D, Almeida A, Biscoito M, Comte T, Khripounoff A, Le Bris N, Sarradin P-M, Segonzac M (2000) A review of the distribution of hydrothermal vent communities along the northern Mid-Atlantic Ridge: dispersal vs. environmental controls. Hydrobiologia 440:201–216. https://doi.org/10.1023/A:1004175211848

Desbruyères D, Biscoito M, Caprais J-C, Colaço A, Comtet T, Crassous P, Fouquet Y, Khripounoff A, Le Bris N, Olu K (2001) Variations in deep-sea hydrothermal vent communities on the Mid-Atlantic Ridge near the Azores plateau. Deep Sea Res Part I Oceanogr Res Pap 48:1325–1346

Erickson K, Macko S, Van Dover C (2009) Evidence for a chemoautotrophically based food web at inactive hydrothermal vents (Manus Basin). Deep Sea Res Part II Top Stud Oceanogr 56:1577–1585

Fabri M-C, Bargain A, Briand P, Gebruk A, Fouquet Y, Morineaux M, Desbruyères D (2011) The hydrothermal vent community of a new deep-sea field, Ashadze-1, 12°58′ N on the mid-Atlantic ridge. J Mar Biol Assoc U K 91:1–13

Früh-Green GL, Orcutt BN, Rouméjon S, Lilley MD, Morono Y, Cotterill C, Green S, Escartin J, John BE, Mccaig AM, Ménez B, Schwarzenbach EM, Morgan J, Morgan S, Lang SQ, Schrenk MO, Brazelton WJ, Akizawa N, Boschi C, Hesse K (2018) Magmatism, serpentinization and life: insights through drilling the Atlantis Massif (IODP expedition 357). Lithos 323:137–155. https://doi.org/10.1016/j.lithos.2018.09.012

Galkin SV (2006) Prostranstvennaya struktura gidrotermalnykh soobshchestv Sredinno-Atlanticheskogo Khrebta. In: Vinogradov ME, Vereshchaka AL (eds) Ekosistemy atlanticheskikh gidroterm. Nauka, Moskva, pp 163–202

Galkin SV, Sagalevich AM (2016) Extreme biomimetics. In: Ehrlich H (ed) Extreme biomimetics. Springer, pp 97–118. https://doi.org/10.1007/978-3-319-45340-8

Gebruk AV, Galkin SV, Vereshchaka AL, Moskalev LI, Southward AJ (1997) Ecology and biogeography of the hydrothermal vent fauna of the Mid-Atlantic Ridge. Adv Mar Biol 32:93–144

Gebruk A, Southward E, Kennedy H, Southward A (2000) Food sources, behaviour, and distribution of hydrothermal vent shrimps at the Mid-Atlantic Ridge. J Mar Biol Assoc U K 80:485–499

Gerdes K, Martínez Arbizu P, Schwarz-Schampera U, Schentner M, Kihara TC (2019) Detailed mapping of hydrothermal vent Fauna: a 3D reconstruction approach based on video imagery. Front Mar Sci 6:96

Glein CR, Baross JA, Waite JH (2015) The pH of Enceladus' ocean. Geochim Cosmochim Acta 162:202–219. https://doi.org/10.1016/j.gca.2015.04.017

Glover AG, Gooday AJ, Bailey DM, Billett DSM, Chevaldonne P, Colaço A, Copley J, Cuvelier D, Desbruyères D, Kalogeropoulou V, Klages M, Lampadariou N, Lejeusne D, Mestre NC, Paterson GLJ, Perez T, Ruhl H, Sarrazin J, Soltwedel T, Soto EH, Thatje S, Tselepides A, Van Gaever S, Vanreusel A (2010) Temporal chnage in deep-sea benthic ecosystmes: a review of the evidence from recent time-searies studies. Adv Mar Biol 58:1–95

Gollner S, Govenar B, Fisher CR, Bright M (2015) Size matters at deep-sea hydrothermal vents: different diversity and habitat fidelity patterns of meio- and macrofauna. Mar Ecol Prog Ser 520:57–66. https://doi.org/10.3354/meps11078

Hilário A, Metaxas A, Gaudron SM, Howell KL, Mercier A, Mestre NC, Ross RE, Thurnherr AM, Young C (2015) Estimating dispersal distance in the deep sea: challenges and applications to marine reserves. Front Mar Sci 2:6. https://doi.org/10.3389/fmars.2015.00006

Hoehler TM (2007) An energy balance concept for habitability. Astrobiology 7:824–838. https://doi.org/10.1089/ast.2006.0095

Jollivet D (1996) Specific and genetic diversity at deep-sea hydrothermal vents: an overview. Biodivers Conserv 5:1619–1653. https://doi.org/10.1007/BF00052119

Kargel JS, Kaye JZ, Head JW, Marion GM, Sassen R, Crowley JK, Grant SA, Hogenboom DL (2000) Europa's crust and ocean: origin, composition, and the prospects for life. Icarus 265:226–265. https://doi.org/10.1006/icar.2000.6471

Kelley DS, Karson JA, Früh GL, Yoerger DR, Shank TM, Butterfield DA, Hayes JM, Schrenk MO, Olson EJ, Proskurowski G, Jakuba M, Bradley A, Larson B, Ludwig K, Glickson D, Buckman K, Bradley AS, Brazelton WJ, Roe K, Sylva SP (2005) A serpentinite-hosted

ecosystem: the Lost City hydrothermal field. Science 307:1428–1434. https://doi.org/10.1126/science.1102556
Lang SQ, Brazelton WJ (2020) Habitability of the marine serpentinite subsurface: a case study of the Lost City hydrothermal field. Philos Trans R Soc A 378:20180429. https://doi.org/10.1098/rsta.2018.0429
Lang SQ, Butterfield DA, Schulte M, Kelley DS, Lilley MD (2010) Elevated concentrations of formate, acetate and dissolved organic carbon found at the Lost City hydrothermal field. Geochim Cosmochim Acta 74:941–952. https://doi.org/10.1016/j.gca.2009.10.045
Lang SQ, Früh-Green GL, Bernasconi SM, Lilley MD, Proskurowski G, Méhay S, Butterfield DA (2012) Microbial utilization of abiogenic carbon and hydrogen in a serpentinite-hosted system. Geochim Cosmochim Acta 92:82–99
Lang SQ, Früh-Green GL, Bernasconi SM, Brazelton WJ, Schrenk MO, Mcgonigle JM (2018) Deeply-sourced formate fuels sulfate reducers but not methanogens at Lost City hydrothermal field. Nat Sci Rep 8:755. https://doi.org/10.1038/s41598-017-19002-5
Larsen K, Błażewicz-Paszkowycz M, Cunha MR (2006) Tanaidacean (Crustacea: Peracarida) fauna from chemically reduced habitats - the Lucky Strike hydrothermal vent system, Mid-Atlantic Ridge. Zootaxa 1187:1–36
Lartaud F, Little CT, De Rafelis M, Bayon G, Dyment J, Ildefonse B, Gressier V, Fouquet Y, Gaill F, Le Bris N (2011) Fossil evidence for serpentinization fluids fueling chemosynthetic assemblages. Proc Natl Acad Sci 108:7698–7703
Le Bris N, Yücel MAD, Sievert SM, LokaBharathi P, Girguis PR (2019) Hydrothermal energy transfer and organic carbon production at the deep seafloor. Front Mar Sci 5:531
Lunina AA, Vereshchaka AL (2010) A new vent shrimp (Crustacea: Decapoda: Alvinocarididae) from the Mid-Atlantic Ridge. Zootaxa 2372:69–74
Martin W, Baross J, Kelley D, Russell MJ (2008) Hydrothermal vents and the origin of life. Nat Rev Microbiol 6:805–814. https://doi.org/10.1038/nrmicro1991
Mason OU, Nakagawa T, Rosner M, van Nostrand JD, Zhou J, Maruyama A, Fisk MR, Giovannoni SJ (2010) First investigation of the microbiology of the deepest layer of ocean crust. PLoS One 5:e15399. https://doi.org/10.1371/journal.pone.0015399
Matabos M, Cuvelier D, Brouard J, Shillito B, Ravaux J, Zbinden M, Barthelemy D, Sarradin P-M, Sarrazin J (2015) Behavioural study of two hydrothermal crustacean decapods: *Mirocaris fortunata* and *Segonzacia mesatlantica*, from the Lucky Strike vent field (Mid-Atlantic Ridge). Deep Sea Res Part II Top Stud Oceanogr 121:146–158
McCollom TM (2007) Geochemical constraints on sources of metabolic energy for chemolithoautotrophy in ultramafic-hosted deep-sea hydrothermal systems. Astrobiology 7:933–950
McCollom TM, Seewald JS (2007) Abiotic synthesis of organic compounds in deep-sea hydrothermal environments. Chem Rev 107:382–401
McCollom TM, Shock EL (1997) Geochemical constraints on chemolithoautotrophic metabolism by microorganisms in seafloor hydrothermal systems. Geochim Cosmochim Acta 61:4375–4391
McDermott JM, Seewald JS, German CR, Sylva SP (2015) Pathways for abiotic organic synthesis at submarine hydrothermal fields. Proc Natl Acad Sci 112:7668–7672. https://doi.org/10.1073/pnas.1506295112
McGonigle JM, Lang SQ, Brazelton WJ (2020) Genomic evidence for formate metabolism by Chloroflexi as the key to unlocking deep carbon in Lost City microbial ecosystems. Appl Environ Microbiol 86:e02583-19. https://doi.org/10.1128/AEM.02583-19
Mironov AN, Gebruk AV, Moskalev LI (2002) Geografia gidrotermalnykh soobshchestv i obligatnykh gidrotermalnykh taksonov. In: Gebruk AV (ed) Biology of hydrothermal systems. KMK Press, Moscow, pp 410–455
O'Mullan GD, Maas PA, Lutz RA, Vrijenhoek RC (2001) A hybrid zone between hydrothermal vent mussels (Bivalvia: Mytilidae) from the Mid-Atlantic Ridge. Mol Ecol 10:2819–2831
Osenberg CW, Schmitt RJ, Holbrook SJ, Abu-Saba KE, Flegal AR (1994) Detection of environmental impacts: natural variability, effect size, and power analysis. Ecol Appl 4:16–30

Oze C, Mousis O, Waite JH (2015) Serpentinization and the formation of H_2 and CH_4 on celestial bodies (planets, moons, comets). Astrobiology 15:587–600. https://doi.org/10.1089/ast.2014.1188
Portail M, Brandily C, Cathalot C, Colaço A, Gélinas Y, Husson B, Sarradin P-M, Sarrazin J (2018) Food-web complexity across hydrothermal vents on the Azores Triple Junction. Deep Sea Res I Oceanogr Res Pap 131:101–120
Proskurowski G, Lilley MD, Kelley DS, Olson EJ (2006) Low temperature volatile production at the Lost City hydrothermal field, evidence from a hydrogen stable isotope geothermometer. Chem Geol 229:331–343. https://doi.org/10.1016/j.chemgeo.2005.11.005
Proskurowski G, Lilley MD, Seewald JS, Früh-Green GL, Olson EJ, Lupton JE, Sylva SP, Kelley DS (2013) Abiogenic hydrocarbon production at Lost City hydrothermal field. Science 319:604–608
Radziejewska T (2014) Meiobenthos in the sub-equatorial Pacific Abyss. A proxy in anthropogenic impact evaluation. Springer, Heidelberg
Reid WD, Wigham BD, McGill RA, Polunin NV (2012) Elucidating trophic pathways in benthic deep-sea assemblages of the Mid-Atlantic Ridge north and south of the Charlie-Gibbs Fracture Zone. Mar Ecol Prog Ser 463:89–103
Roth SE, Dymond J (1989) Transport and settling of organic material in a deep-sea hydrothermal plume: evidence from particle flux measurements. Deep Sea Res Part I Oceanogr Res Pap 36:1237–1254
Rybakova E, Galkin S (2015) Hydrothermal assemblages associated with different foundation species on the East Pacific Rise and Mid-Atlantic Ridge, with a special focus on mytilids. Mar Ecol 36(Suppl. 1):45–61
Sarrazin J, Cuvelier D, Peton L, Legendre P, Sarradin P-M (2014) High-resolution dynamics of a deep-sea hydrothermal mussel assemblage monitored by the EMSO-Açores MoMAR observatory. Deep Sea Res Part I Oceanogr Res Pap 90:62–75
Schrenk MO, Kelley DS, Bolton SA, Baross JA (2004) Low archaeal diversity linked to subseafloor geochemical processes at the Lost City hydrothermal field, Mid-Atlantic Ridge. Environ Microbiol 6:1086–1095. https://doi.org/10.1111/j.1462-2920.2004.00650.x
Segonzac M (1992) Les peuplements associés a l'hydrothérmalisme océanique du Snake Pit (dorsal Médio-Atlantique; 23°N, 3480 m): Composition et microdistribution de la megafaune. C R Acad Sci 314:593–600
Silva MAA (2016) Assessing potential impacts of deep-sea mining on dispersal and population connectivity of the vent mussel *Bathymodiolus azoricus*. Ph.D. thesis, Universidade dos Açores
Sleep NH, Meibom A, Fridriksson T, Coleman RG, Bird DK (2004) H_2-rich fluids from serpentinization: geochemical and biotic implications. Proc Natl Acad Sci 101:12818–12823. www.pnas.org/cgi/doi/10.1073/pnas.0405289101
Stüeken EE, Anderson RE, Bowman JS, Brazelton WJ, Colangelo-Lillis J, Goldman AD, Som SM, Baross JA (2013) Did life originate from a global chemical reactor? Geobiology 11:101–126. https://doi.org/10.1111/gbi.12025
Summit M, Baross JA (2001) A novel microbial habitat in the mid-ocean ridge subseafloor. Proc Natl Acad Sci 98:2158–2163
Sutton T, Porteiro F, Heino M, Byrkjedal I, Langhelle G, Anderson C, Horne J, Søiland H, Falkenhaug T, Godø O (2008) Vertical structure, biomass and topographic association of deep-pelagic fishes in relation to a mid-ocean ridge system. Deep Sea Res Part II Top Stud Oceanogr 55:161–184
Teixeira S, Cambon-Bonavita MA, Serrão EA, Desbruyéres D, Arnaud-Haond S (2011) Recent population expansion and connectivity in the hydrothermal shrimp *Rimicaris exoculata* along the Mid-Atlantic Ridge. J Biogeogr 38:564–574. https://doi.org/10.1111/j.1365-2699.2010.02408.x
Van Dover C (2000) The ecology of deep-sea hydrothermal vents. Princeton University Press
Van Dover CL (2014) Impacts of anthropogenic disturbances at deep-sea hydrothermal vent ecosystems: a review. Mar Environ Res 102:59–72

Van Dover CL, Doerries MB (2005) Community structure in mussel beds at Logatchev hydrothermal vents and a comparison of macrofaunal species richness on slow- and fast-spreading mid-ocean ridges. Mar Ecol 26:110–120
Vereshchaka A, Vinogradov G, Lein AY, Dalton S, Dehairs F (2000) Carbon and nitrogen isotopic composition of the fauna from the Broken Spur hydrothermal vent field. Mar Biol 136:11–17
Von Damm KL (2001) Lost City found. Nature 412:127–128
Waite JH, Glein CR, Perryman RS, Teolis BD, Magee BA, Miller G, Grimes J, Perry ME, Miller KE, Bouquet A, Lunine JI, Brockwell T, Bolton SJ (2017) Cassini finds molecular hydrogen in the Enceladus plume: evidence for hydrothermal processes. Science 159:155–159
Yishai O, Lindner SN, Gonzalez J, Cruz D, Tenenboim H, Bar-Even A (2016) The formate bio-economy. Curr Opin Chem Biol 35:1–9. https://doi.org/10.1016/j.cbpa.2016.07.005

Part III
Research Methods

Deep Sea Research Methods to Be Used During the Exploration of the Mid-Atlantic Ridge Polymetallic Sulphide Areas

Teresa Radziejewska, Jan Marcin Węsławski, Michał Tomczak, Sławomir Sagan, and Łukasz Smajdor

1 Water Column Ecosystem, Imagery

Deep sea research and monitoring belong to the most expensive and technically challenging marine operations. The modern approach focuses on applying non-invasive and non-destructive methods, so all types of imagery and recording are in use, from microscopic *in situ* photography of the water column to detect marine snow and the microplankton, to acoustics, both passive (listening to sounds of marine organisms) and active (recognition of objects and structures in the water column and on the seabed). Some destructive sampling is unavoidable (plankton, fish, small nekton), the appropriate gear being listed in Table 1. The seabirds and marine mammal censuses involve both direct sightings and counting from the ship (the Tasker method) and passive acoustic buoys recording vocalization of marine mammals.

Although imagery is a key technique for the benthic biota—from small scale photoframes (cm^2) to larger seascape views (m^2), it has also been used to survey pelagic communities. For example, Vereshchaka and Vinogradov (1999) as well as Vinogradov et al. (2003), while working from a manned submersible (= human operated vehicle, HOV) combined direct observations, underwater photography,

T. Radziejewska
Institute of Marine and Environmental Sciences, University of Szczecin, Szczecin, Poland
e-mail: teresa.radziejewska@usz.edu.pl

J. M. Węsławski (✉) · S. Sagan
Institute of Oceanology PAN, Sopot, Poland
e-mail: weslaw@iopan.gda.pl; sagan@iopan.gda.pl

M. Tomczak · Ł. Smajdor
Polish Geological Institute – National Research Institute, Warsaw, Poland
e-mail: mtomc@pgi.gov.pl

T. Radziejewska et al. (eds.), *The Natural Environment of the Mid-Atlantic Ridge*, GeoPlanet: Earth and Planetary Sciences,
https://doi.org/10.1007/978-3-031-51865-2_7

Table 1 Types of gear recommended for the deep sea exploration

Component observed	Main tools	Remarks	References
CTD profiles	CTD system with rosette water cast	Variety of standard gear	White et al. (2016)
Optical parameters in the epipelagial and at depth	Optical sensors	Mounted on ROV, AUV and landers	Cave and German (1998)
Plankton	Plankton nets, including multinet; direct visual observations from HOV and ROV		Christiansen (2016)
Mesopelagic fish and small nekton	RMT trawl; multinet	Shrimp pelagic trawl can also be used; multinet designs, e.g., MOCNESS and Kiel	Clark et al. (2016a, b), Christiansen (2016)
Sedimentation	Sediment traps deployed for various periods of time	Standard gear	White et al. (2016)
Carnivorous and carrion feeding mobile fauna	Baited traps	Variety of non-standarised traps, size-dependent	Jamieson (2016)
Encrusting and sessile fauna	Photo documentation by ROV operator arms	Restricted destructive sampling	Carroll et al. (2018), Bowden and Jones (2016)
Soft-sediment fauna	Push-corer; multiple corer (multi-corer, MUC); mega-corer; box-corer	Push-corers, operated from ROV or HOV; box corers of different design, e.g., USNEL MkII box corer	Barnett et al. (1984) Bell et al. (2016) Narayanaswamy et al. (2016a, b)
Geochemical processes	Lander with chemical probes	*Ad hoc* constructions and configuration	Jamieson (2016)

and plankton net hauls to sample zooplankton in a number of hydrothermal fields on the MAR (including Broken Spur and Lost City in the Polish contract area).

Description of various gear used in deep-sea exploration, in various compartments of the oceanic realm, can be found in a monograph edited by Clark et al. (2016a, b) and in references therein (indicated also in Table 1). In addition, a good review of the state-of-the-art equipment for deep sea pelagic sampling is provided by the web page describing the gear prepared for a new UK esearch vessel RV Sir Richard Attenborough (https://www.bas.ac.uk/polar-operations/sites-and-facilities/facility/rrs-sir-david-attenborough/science-facilities/net-trawl-sledge-systems/).

2 Optical Properties

The state of the epipelagial of the geographically remote areas can be successfully monitored remotely down to hundreds of meters with a swath of sensors designed for observations of various phenomena and processes. Features such as productivity, chlorophyll and particle content as well as dissolved organic matter can be assessedd with already fully developed and robust algorithms where the water constituencies are parameterised by a validated and calibrated combination of water-leaving radiances of carefully chosen light wavelengths in the visible part of the spectrum (Blondeau-Patissier et al. 2014; Vandermeulen et al. 2020). Other remote sensing methods allow to observe surface currents, waves, wind speed, water mass movements as well as the deep seafloor topography (Sandwell et al. 2014).

Advantageous is the deployment of optical sensors together with acoustic ones and the and CDT, as the particulate enrichment extends much further, on the order of several kilometres, than temperature anomalies. Hydrothermal vents produce plumes spread with the aid of currents, and can be detected at a distance by observing hydro-physical properties of the water. Turbidity anomalies were documented at maximum level of 0.10 ΔNTU (Nephelometric Turbidity Units) (Tao et al. 2017). Additionally, enhanced light scattering by particles might be easily connected with acoustically detected bottom features (Cave and German 1998). In most of the observations, the plume core retains its initial depth (Fig. 1). Other sources of particle flow are the abyssal currents (Lahaye et al. 2019) or the open-ocean convection which may produce a resuspension particle flow of thickness of meters above the seafloor (Madron et al. 2017), reducing the water transparency in the areas affected.

Monitoring of the deep-sea bottom requires demanding and robust techniques and instrumentation, such as remotely operated vehicles (ROVs), autonomous underwater vehicles (AUVs) and bottom landers with optical or acoustic telemetry. It has to be noted that permanent or long-term installations, either floating or located on the seafloor, might be subject to macrofouling, despite the apparent absence of the living material at large depths (Blanco et al. 2013).

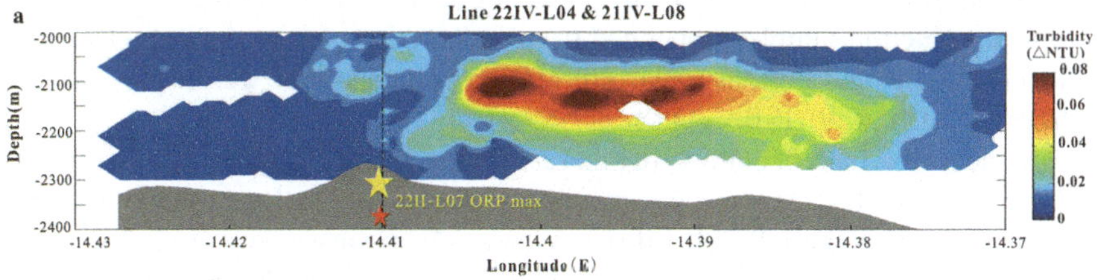

Fig. 1 Transects of turbidity from survey lines spanning 2009 to 2011. 21IV-L04/21IV-L08. Red stars show the vent location (Tao et al. 2017)

3 Benthos

The choice of methods, including sampling gear, to be used for benthic community surveys during the exploration of the Polish contract area on the Mid-Atlantic Ridge (MAR) should primarily follow recommendations of the regulatory body [International Seabed Authority's Legal and Technical Commission (LTC)]; these are contained in the document ISBA/25/LTC/6/Rev.1 (ISBA 2020). In addition, standard practices of deep-sea biological surveys should be adhered to; these have been summarised, for different compartments of the deep-sea realm, in a collection of papers contained in a volume edited by Clark et al. (2016a, b).

The methods and gear will differ depending on the habitat to be surveyed (rocks, soft sediments) and on the actual benthos category to be studied (mega-, macro-, meio- and microbenthos).

Data on the abundance, biomass, species structure and diversity of the megafauna are preferably collected with underwater video and photography (with a scale visible, e.g., using lasers set a defined distance apart), taken with towed cameras (Bowden and Jones 2016), preferably from a remotely operated vehicle (ROV) (Kelley et al. 2016). Recently, substantial technological advances have been made with respect to the image resolution, particularly when the images (still and video) are recorded from ROVs with real-time shipboard control, which makes it possible to zoom in on the required details, even those of individual organisms (cf. photos in Fig. 2). Advances have also been made in the annotation software (Gomes-Pereira et al. 2016), whereby organisms observed and photographed can be identified in real time, with the assistance of and reference to both taxonomic expertise and on-line image repositories. Landers with time-lapse cameras (Jamieson 2016), deployed for periods of up to several months, have been used to document temporal changes in seafloor communities. However, they have been relatively more popular in abyssal plains than at hydrothermal fields.

It is important that photographs have sufficient resolution to identify organisms larger than 2 cm in the smallest dimension. The seafloor area photographed should be at least 2 m across. It is also important that the images are georeferenced. Species identification should be confirmed by collecting specimens at the site, e.g., by using slurp guns (suction samplers) operated from an ROV (Copley et al. 2016). Dredges, epibenthic sleds and bottom trawls are not recommended for use in polymetallic sulphide (PMS) areas (ISBA 2020).

The deep-sea macrobenthos (defined as proto- and metazoans retained on the 300 μm sieve mesh size) is usually sampled, from the soft-sediment seafloor, with a box corer, such as the USNEL MkII box-corer (Narayanaswamy et al. 2016a, b). Other corers can be deployed, such as a Bowers and Connelly dampened megacorer (Gage and Bett 2005) used by Bell et al. (2016) to sample the macrofauna from edifices on sedimented hydrothermal vent sites at the Bransfield Strait (Antarctica). Alternatively, push-corers of an appropriate diameter, operated by an ROV (Kelley et al. 2016) should be retrieved from soft-sediment areas. Sampling the macrofauna (and also the meiobenthos) associated with megafaunal assemblages on PMS hard

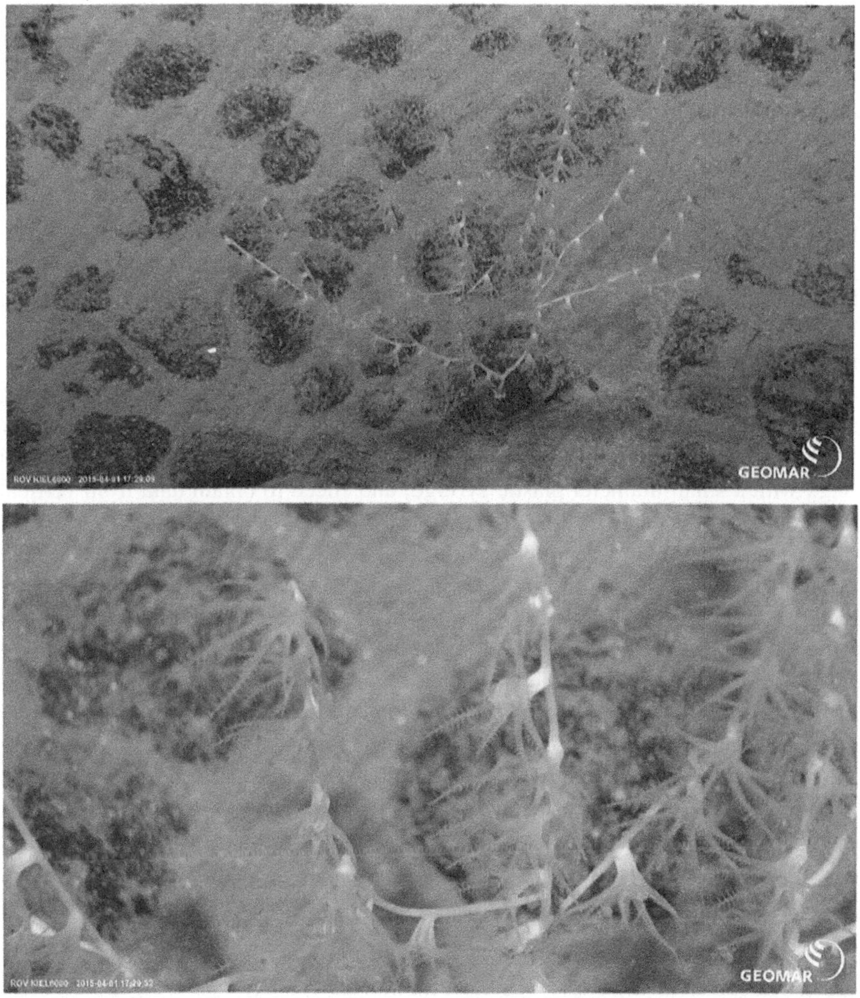

Fig. 2 Illustration of the level of morphological detail achievable with recent ROV-borne photo imagery; a hydroid attached to a polymetallic nodule in the Clarion-Clipperton Zone (Pacific); Cruise SO 239, RV Sonne (Photo credit: GEOMAR, Kiel)

(rock) surfaces is challenging, and has rarely been accomplished, although highly desirable (S. Gollner, pers. comm.), also in view of the ISA recommendations for obtaining data from different sub-habitats (ISBA 2020). For such purposes, Gollner et al. (2006) used a quantitative 'pot' sampler (as described and illustrated in Van Dover 2002) and a 'Bushmaster Jr.' device (Govenar et al. 2005) handled from a human-operated vehicle (HOV). The Bushmaster Jr. is lined with a 63 μm Nitex mesh and can sample areas up to 60 cm in diameter (0.28 m^2).

The abundance, biomass and taxonomic structure of the protozoan (foraminiferal) and metazoan meiofauna (organisms retained on the 32 μm sieve mesh size) should be determined from quantitative samples. These are usually collected with a multiple corer (multi-corer, MUC) such as an SMBA MUC (Barnett et al. 1984; Narayanaswamy et al. 2016a, b). Recent meiofauna studies on MAR (Spedicato et al. 2020), however, involved push-corers (5.5 cm diameter) operated from an ROV (Flint 2007; cf. Fig. 3), and these seem to be a sampling gear of choice, also on account of providing sediment samples for the study of microbiota (see below). After retrieval, the cores, regardless of their type, should be sliced into 1-cm thick layers down to 5 cm (ISBA 2020). As pointed out above, of interest in the PMS setting would be to collect meiofauna associated with the megafauna assemblages on hard substrates. For such purposes, the sampling gear—mentioned above—applied by Gollner et al. (2006), but requiring the use of an HOV rather than an ROV, can be applicable.

The microbial community, important for studying the ecosystem functioning, should be sampled—from soft sediments—using any corer (a MUC, a push-corer from an ROV). Usually, cores obtained with a MUC or a push-corer are subsampled (e.g., with a small-diameter plastic syringe) and sliced to obtain data from vertical intervals 1-cm thick (ISBA 2020).

As stated by the ISA recommendations (ISBA 2020), "standard practices for the fixation and preservation of organisms should be followed, while noting that formaldehyde/formalin fixation is not appropriate for all taxa". Depending on a taxon and study objective, multiple preservation methods may be necessary, including freezing or preservation in molecular grade ethanol for molecular studies; preservation in ethanol or formaldehyde for morphological taxonomic studies; or the immediate freezing of whole animals and/or selected tissues for stable isotope, trace metal and biochemical analyses (Schiaparelli et al. 2016). In the latter context, Riehl et al. (2014) described methods of tissue sample collection for DNA studies on deep-sea crustaceans they had successfully applied in their studies.

Fig. 3 Push-cores inserted in the sediment (Photo credit: GEOMAR, Kiel)

4 Geological Structures

Conventional marine-geological surveys involve ground-truthing methods which provide geological samples as well as physical evidence of seabed structure, usually at a local scale. Seafloor sediments can be sampled with a range of tools, and the collected samples are used to determine geophysical and geochemical parameters useful in deciphering the driving forces undelying sediment formation and post-depositional processes affecting sediment properties. The spatial resolution of such samples is usually fairly low, and sampling operations are complex, costly, and time-consuming. Therefore, these methods are appropriate for identification or characterisation of small-scale structures.

However, regional assessments need a more general picture of the sediment structures. Increasing the sampling coverage can deliver information about large area, but the final results are based on interpolation between the sampling points, which carries some level of uncertainty. Such sampling approach is frequently applied following a regional reconnaissance achieved by a continued acquisition of geophysical data, i.e., the bathymetric survey achieved with a Multibeam Echo Sounder System (MBES) (e.g., Machida et al. 2019). In some cases, the actual sediment samples and their properties may provide information necessary for the correct calibration of certain geophysical methods.

The most important geophysical methods used for structural geology interpretation are those related to the acoustic wave propagation and density measurements by a multibeam echosounder, a sub-bottom profiler as well as seismic and gravimetric equipment, and measurements of magnetic and electric fields. The level of detail obtained with these methods is directly related to the survey design.

The surveys are carried out on different spatial scales: regional, semi-detailed, and detailed. The methods and the equipment used will depend on the level of detail required, which in turn will be achieved depending on the distance to the seafloor (altitude above the sea floor) the survey operation is being carried out.

Regional surveys provide information on large areas (on the order of 10^2–10^3 km) and involve the use of equipment necessary for ship-borne bathymetry (MBES), ship-borne side-scan sonar (SSS), sub-bottom profiling (SBP), as well as the equipment with which to carry out seismic wave (reflection from boundaries having different acoustic impedance, appropriate for detecting large structures) and ship-borne gravity and magnetic field (to detect density and gravity anomalies of underlying rocks) surveys (Table 2) (modified after JAMSTEC 2018).

Semi-detailed surveys provide high-resolution (both lateral and vertical) information on small areas (on the order of several km^2) and involve the use of AUV/ROV-borne MBES and AUV-borne SBP as well as the equipment with which to carry out autonomous cabled seismic (ACS) reflection (high lateral and vertical resolution) and AUV-borne gravity and magnetic surveys (Table 2).

Detailed surveys supply very high-resolution information on small spatial scale areas (a metre-scale resolution). The methods used include Vertical Cable Seismic (VCS) reflection, with hydrophone arrays; Zero-Offset VSC (ZVCS), used in oil

Table 2 Methods and equipment used for seafloor mapping and exploration

Survey scale	Exploration technique	Survey component	Platform
Regional	Acoustic	MBES bathymetry Sub-bottom profiler Seismic reflection survey Side Scan Sonar	Ship
	Gravitmetry and mag	Gravity and magnetic survey	
Semi-detailed	Acoustic	AUV-based survey (MBES, SSS etc.) Seismic reflection by ACS	Deep-tow, AUV
	Gravitmetry and magnetometry	AUV-borne gravity and magnetic survey	
Detailed	Acoustic	Seismic reflection by VCS SAS/SSS Interferometric Bathymetry Sub-bottom profiler	ROV and AUV
	Gravitmetry and magnetometry	ROV based gravity and magnetic survey	
	Geological survey (sediment sampling)	Dredge Grab Box-corer Multi-corer Gravity-corer ROV/ROV, Crawler Drilling rig	Ship-borne, Seafloor-based operations, ROV

and gas exploration (Asakawa et al. 2016, 2019), ROV-borne gravity and magnetic surveys, Synthethic Aperture Sonar (SAS) surveys (resolution 3–5 cm) (Hansen 2013; Kowalczyk and Lum 2017), and AUV-borne interferometric bathymetry surveys (resolution of 2–10 cm) (Kowalczyk and Lum 2017).

The surveys described above are complemented with point-source information on the sediment/deposit structure (including mineral composition), obtained from actual deposit samples collected with dredges (triangular or rectangular, with bottom-setting or target-setting design); solid deposit grabs (frequently equipped with cameras to document the process of sampling); box corers and multiple corers (multi-corers) for sampling soft sediments, providing cores of undisturbed sediment; gravity corers to obtain cores of up to 12 m in length, and providing data and insights on time-course of sediment accumulation on very long temporal scales.

The methods and gear listed above are deployed from a variety of platforms (cf. Table 2): research vessels, remotely operated vehicles (ROV), autonomously operated vehicles (AOV), and drilling rigs, of which the first three constitute standard equipment used during exploration on mid-ocean ridges, including the Mid-Atlantic Ridge (MAR).

References

Asakawa E, Murakami F, Tsukahara H, Saito S, Lee S, Tara K (2016) New marine seismic survey techniques for seafloor massive sulphide (SMS) exploration. In: Proceedings of IEEE 2016 Techno-Ocean 2016, Kobe, Japan. Curran Associates, pp 134–139

Asakawa E, Lee S, Sumi T, Endo M (2019) Integrated geophysical approach for the seafloor massive sulfide (SMS) exploration. https://www.essoar.org/pdfjs/10.1002/essoar.10500412.1

Barnett PRO, Watson J, Conelly D (1984) A multiple corer for taking virtually undisturbed samples from shelf, bathyal and abyssal sediments. Oceanol Acta 7:399–408

Bell JB, Woulds C, Brown LE, Sweeting CJ, Reid WDK, Little CTS, Glover AG (2016) Macrofaunal ecology of sedimented hydrothermal vents in the Bransfield Strait, Antarctica. Front Mar Sci 3:32. https://doi.org/10.3389/fmars.2016.00032

Blanco R, Shields MA, Jamieson AJ (2013) Macrofouling of deep-sea instrumentation after three years at 3690 m depth in the Charlie Gibbs Fracture Zone, Mid-Atlantic Ridge, with emphasis on hydroids (Cnidaria: Hydrozoa). Deep-Sea Res Part II Top Stud Oceanogr 98:370–373. https://doi.org/10.1016/j.dsr2.2013.01.019

Blondeau-Patissier D, Gower JFR, Dekker AG, Phinn SR, Brando VE (2014) A review of ocean color remote sensing methods and statistical techniques for the detection, mapping and analysis of phytoplankton blooms in coastal and open oceans. Prog Oceanogr 123:123–144. https://doi.org/10.1016/j.pocean.2013.12.008

Bowden D, Jones D (2016) Towed cameras. In: Clark M, Consalvey M, Rowden AA (eds) Biological sampling in the deep sea. Wiley Blackwell, Chichester, pp 260–284

Carroll A, Althaus F, Beaman R, Friedman A, Ierodiaconou D, Ingleton T, Jordan A, Linklater M, Monk J, Post A, Przeslawski R, Smith J, Stowar M, Tran M, Tyndall A (2018) Marine sampling field manual for towed underwater camera systems. In: Przeslawski R, Foster S (eds). Field manuals for marine sampling to monitor Australian Waters, National Environmental Science Programme (NESP), pp 131–152

Cave RR, German CR (1998) Hydrothermal plume detection in the deep ocean - a combination of technologies. Underw Technol 23:71–75. https://doi.org/10.3723/175605498783258921

Christiansen B (2016) Deep-sea zooplankton sampling. In: Clark M, Consalvey M, Rowden AA (eds) 2016. Biological sampling in the deep sea, Wiley Blackwell, Chichester, pp 103–125

Clark M, Bagley N, Harley B (2016a) Trawls. In: Clark M, Consalvey M, Rowden AA (eds) 2016. Biological sampling in the deep sea, Wiley Blackwell, Chichester, pp 126–158

Clark M, Consalvey M, Rowden AA (eds) (2016b) Biological sampling in the deep sea. Wiley Blackwell, Chichester

Copley JT, Marsh L, Glover AG, Hühnerbach V, Nye VE, Reid WDK, Sweeting CJ, Wigham BD, Wiklund H (2016) Ecology and biogeography of megafauna and macrofauna at the first known deep-sea hydrothermal vents on the ultraslow-spreading Southwest Indian Ridge. Sci Rep 6:39158

Flint HC (2007) Diversity of meiofauna at deep-sea hydrothermal vents and cold seeps with particular reference to nematodes. Ph.D. Thesis, National Oceanography Centre, Southampton, UK

Gage JD, Bett BJ (2005) Deep-sea benthic sampling. In: Holmes NA, McIntyre AD (eds) Methods for the study of the marine benthos. 3rd edn. Blackwell Scientific Publications, Oxford. https://doi.org/10.1002/9781118542392

Gollner S, Zekely J, Van Dover CL, Govenar B, Le Bris N, Nemeschka HL, Bright M (2006) Benthic copepod communities associated with tubeworm and mussel aggregations on the East Pacific Rise. Cah Biol Mar 47:397–402

Gomes-Pereira JN, Auger V, Beisiegel K, Benjamin R, Bergmann M, Bowden D, Buhl-Mortensen P, De Leo FC, Dionísio G, Durden JM, Edwards L, Friedman A, Greinert J, Jacobsen-Stout N, Lerner S, Leslie M, Nattkemper TW, Sameoto JA, Schoening T, Schouten R, Seager J, Singh H, Soubigou O, Tojeira I, van den Beld I, Dias F, Tempera F, Santos RS (2016) Current and future trends in marine image annotation software. Prog Oceanogr 149:106–120

Govenar B, Le Bris N, Gollner S, Glanville J, Aperghis A, Hourdez S, Fisher CR (2005) Epifaunal community structure associated with *Riftia pachyptila* aggregations in chemically different hydrothermal vent habitats. Mar Ecol Prog Ser 305:67–77

Hansen RE (2013) Synthetic aperture sonar technology review. Mar Technol Soc J 47:117–127

ISBA (2020) Recommendations for the guidance of contractors for the assessment of the possible environmental impacts arising from exploration for marine minerals in the area issued by the legal and technical commission. ISBA/25/LTC/6/Rev.1

Jamieson A (2016) Landers: baited cameras and traps. In: Clark M, Consalvey M, Rowden AA (eds) Biological sampling in the deep sea. Wiley Blackwell, Chichester, pp 228–259

JAMSTEC (2018) Cross-ministerial strategic innovation promotion program. Next-generation technology for ocean resources exploration (Zipangu in the Ocean Program). Japan Agency for Marine-Earth Science and Technology Project Team for Development of New-generation Research Protocol for Submarine Resources, Yokosuka

Kelley C, Kerby T, Sarradin P-M, Sarrazin J, Lindsay DJ (2016) Submersibles and remotely operated vehicles. In: Clark MR, Consalvey M, Rowden AA (eds) Biological sampling in the deep sea. Wiley Blackwell, Chichester, pp 285–305

Kowalczyk P, Lum B (2017) Sea floor mining exploration technology and methods. In: Tshirhart B, Thomas MD (eds) Proceedings of Exploration 17: Sixth Decennial International Conference on Mineral Exploration, Toronto, Canada, pp 305–318

Lahaye N, Gula J, Thurnherr AM, Reverdin G, Bouruet-Aubertot P, Roullet G (2019) Deep currents in the rift valley of the north Mid-Atlantic Ridge. Front Mar Sci 6:597. https://doi.org/10.3389/fmars.2019.00597

Machida S, Sato T, Yasukawa K, Nakamura K, Iijima K, Nozaki T, Kato Y (2019) Visualisation method for the broad distribution of seafloor ferromanganese deposits. Mar Georesour Geotechnol 39:267–279. https://doi.org/10.1080/1064119X.2019.1696432

Madron XDD, Ramondenc S, Berline L, Houpert L, Bosse A, Martini S, Guidi L, Conan P, Curtil C, Delsaut N, Kunesch S, Ghiglione J, Marsaleix P, Pujo-Pay M, Séverin T, Testor P, Tamburini C (2017) Deep sediment resuspension and thick nepheloid layer generation by open-ocean convection. J Geophys Res Oceans 122:2291–2318. https://doi.org/10.1002/2016JC012062

Narayanaswamy B, Bett B, Lamont P, Rowden A, Bell E, Menot L (2016a) Corers and grabs. In: Clark M, Consalvey M, Rowden AA (eds) Biological sampling in the deep sea. Wiley Blackwell, Chichester, pp 207–259

Narayanaswamy BE, Bett BJ, Lamont PA, Rowden AA, Bell EM, Menot L (2016b) Corers and grabs. In: Clark MR, Consalvey M, Rowden AA (eds) Biological sampling in the deep sea. Wiley Blackwell, Chichester, pp 207–227

Riehl T, Brenke N, Brix S, Driskell A, Kaiser S, Brandt A (2014) Field and laboratory methods for DNA studies on deep-sea isopod crustaceans. Pol Polar Res 35:203–224. https://doi.org/10.2478/popore-2014-0018

Sandwell DT, Müller RD, Smith WHF, Garcia E, Francis R (2014) New global marine gravity model from CryoSat-2 and Jason-1 reveals buried tectonic structure. Science 346:65–67. https://doi.org/10.1126/science.1258213

Schiaparelli S, Schnabel K, Richer de Forges B, Chan T-Y (2016) Sorting, recording, preservation and storage of biological samples. In: Clark M, Consalvey M, Rowden AA (eds) Biological sampling in the deep sea. Wiley Blackwell, Chichester, pp 338–367

Spedicato A, Sánchez N, Pastor L, Menot L, Zeppilli D (2020) Meiofauna community in soft sediments at TAG and Snake Pit hydrothermal vent fields. Front Mar Sci 7:200. https://doi.org/10.3389/fmars.2020.00200

Tao C, Chen S, Baker ET, Li H, Liang J, Liao S, Chen YJ, Deng X, Zhang G, Gu C, Wu J (2017) Hydrothermal plume mapping as a prospecting tool for seafloor sulfide deposits: a case study at the Zouyu-1 and Zouyu-2 hydrothermal fields in the southern Mid-Atlantic Ridge. Mar Geophys Res 38:3–16. https://doi.org/10.1007/s11001-016-9275-2

Van Dover CL (2002) Community structure of mussel beds at deep-sea hydrothermal vents. Mar Ecol Prog Ser 230:137–158

Vandermeulen RA, Mannino A, Craig SE, Werdell PJ (2020) 150 shades of green: using the full spectrum of remote sensing reflectance to elucidate color shifts in the ocean. Remote Sens Environ 247:111900. https://doi.org/10.1016/j.rse.2020.111900

Vereshchaka AL, Vinogradov GM (1999) Visual observations of the vertical distribution of plankton throughout the water column above Broken Spur vent field, Mid-Atlantic Ridge. Deep Sea Res Part I Oceanogr Res Pap 46:1615–1632

Vinogradov GM, Vereshchaka AL, Aleinik DL (2003) Zooplankton distribution over hydrothermal fields of the Mid-Atlantic Ridge. Oceanology 43:696–709

White M, Mohn C, Kiriakoulakis K (2016) Environmental sampling. In: Clark M, Consalvey M, Rowden AA (eds) 2016. Biological sampling in the deep sea, Wiley Blackwell, Chichester, pp 57–79

Part IV
Environmental Protection and Conservation: Legal Framework

Deep-Sea Environmental Protection and Conservation Issues as Applied to the MAR Ecosystem

Teresa Radziejewska, Magdalena Błażewicz, Maria Włodarska-Kowalczuk, and Elżbieta Maszloch

1 Ecosystem Resilience and Connectivity

A system's resilience is a trait encompassing three major components: resistance (sometimes referred to as 'sensitivity'), recovery, and reversibility (Goode et al. 2020).

Resistance is defined (Lotze et al. 2011) as the ability of a system to remain unchanged despite disturbances. It can be assessed at a population and community level. During a PMS mining operation, there will be no question of organisms occurring at the mining site to resist being removed. However, at a distance from the mining site, under potential exposure to tailing discharge and other disturbance, resistance may be exhibited by populations and communities capable of withstanding such disturbance, by analogy to seamount populations and communities showing resistance to disturbance imposed by seabed trawling (Goode et al. 2020). However, resistance of populations at disturbed sites may in fact result from another process, whereby such sites are open for colonisation by opportunistic species

T. Radziejewska (✉)
Institute of Marine and Environmental Sciences, University of Szczecin, Szczecin, Poland
e-mail: teresa.radziejewska@usz.edu.pl

M. Błażewicz
Faculty of Biology, University of Łódź, Łódź, Poland
e-mail: magdalena.blazewicz@biol.uni.lodz.pl

M. Włodarska-Kowalczuk
Institute of Oceanology PAN, Sopot, Poland
e-mail: maria@iopan.gda.pl

E. Maszloch
Polish Geological Institute – National Research Institute, Warsaw, Poland

T. Radziejewska et al. (eds.), *The Natural Environment of the Mid-Atlantic Ridge*, GeoPlanet: Earth and Planetary Sciences,
https://doi.org/10.1007/978-3-031-51865-2_8

(Goode et al. 2020). Therefore, the extent of post-disturbance resistance can only be assessed based on knowledge of the pre-disturbance status of the system, with its natural variability (see below).

Recovery is defined (Goode et al. 2020) as an ability of a system to exhibit improvement (in, e.g., abundance, diversity, productivity) following disturbance, ideally towards pre-disturbance levels (Lotze et al. 2011). If happening, recovery may be complete (a full return to the initial, pre-disturbance state) or partial (the post-disturbance state shows alteration and/or reduction of specific characteristics, relative to the pre-disturbance state). Whether or not recovery after a PMS mining operation will occur is open to debate. Observations at active vent sites showed the communities there, when disturbed and depleted, to be able to recover, the recovery taking, however, up to decades at fast-spreading ridges (Gollner et al. 2017). When initially PMS mining was envisaged as occurring at active vents (Clark and Smith 2013), this propensity to recovery was viewed as an asset. However, as stated by Gollner et al. (2017), reversibility of mining impacts at active vents will depend on the frequency and scale of operations. Large-scale mining may considerably reduce populations and could hinder recolonisation by disrupting connectivity of source populations (Van Dover 2014). The loss of foundation species may make it difficult for the associated fauna to settle, and the system may move to an alternative state dominated by bacterial mats instead of rich faunal communities. The loss of topographic features such as black smokers may change settlement behaviour, and thus the diversity of fauna (Mullineaux and Butman 1990). In addition, modification of vent fluid regimes as a result of black smoker removal could cause local habitat loss or change community composition at both active vents and sites not directly affected by mining.

Nowadays, however, the prevailing opinion holds that PMS will be mined at inactive vents the communities of which are much less known (Gollner et al. 2017; Van Dover 2019). It is unclear whether assemblages at inactive vents are similar to those at active ones, are similar to the fauna inhabiting hard substrata on non-mineral-rich grounds, or are unique (Van Dover 2011). It has to be borne in mind that removal of the PMS substratum at inactive vents cannot be reversed due to the lack of active hydrothermal venting. The limited knowledge on species diversity, functional groups, species distribution, connectivity, and settlement behaviour makes it impossible to estimate reversibility potential at inactive vents and in areas peripheral to vents, following mining disturbances. Primary considerations include habitat alterations; it is not known to what extent, if at all, species will be able to settle and grow in modified environments.

Recovery, whether at a mining site or at a distance from it, will depend on recolonisation potential of neighbouring communities, that is the availability of recruits (including the donor population size) and the suitability of the post-disturbance substrate for settlement and growth. Experiments with artificial substrates placed in the vicinity of active venting sites showed colonisation to occur within months (meiofauna: Cuvelier et al. 2014; Zeppilli et al. 2015) to years (megafauna, e.g., gorgonian corals; Lacharité and Metaxas 2013; Girard et al. 2016). As pointed out above, no estimation of recovery potential of inactive site communities is available (Van Dover et al. 2020).

Reversibility is associated, on the one hand, with the severity of disturbance, and with a system's ability to recover on the other (Lotze et al. 2011). A more severe disturbance will extend the time for the system to recover, provided some degree of recovery is possible. However, as pointed out by Lotze et al. (2011), it is difficult to distinguish reversible changes from alternative stable states a system may enter, or from phase shifts induced by various drivers, including climate change (Levin et al. 2020). With reference to PMS mining, whether the disturbance will produce reversible or irreversible damage will depend on a number of factors, including the disturbance severity and site. The system's ability to recover (and hence reversibility of changes and overall resilience) may also depend on connectivity between the communities affected by disturbance, whereby a population at a defaunated site can be re-established by recolonisation.

Connectivity is a term reflecting ecological connections between populations living distances apart, such as those at hydrothermal vents that are patchy and transient, and is effected primarily via larval dispersal mechanisms (Vrijenhoek 2010; Baldrighi et al. 2018). As experimentally demonstrated by Yearsley et al. (2020), the spatial scale of dispersal varies along the ridge axis, with median dispersal distances for planktonic larval durations (PLD) of 75 days ranging from 67 km to 304 km, which leads to considerable opportunities for connectivity through mid-water dispersal. A stable pattern of five regions of biophysical connectivity was obtained for PLDs of 100 days or more. Connectivity barriers between these regions can persist even when planktonic larval duration is longer than 200 days. For a 50-day PLD, one connectivity barrier coincides with the region of the genetic hybrid zone for northern and southern vent mussel species at the Broken Spur vent field (in the Polish contract area). However, estimates of connectivity between populations of invertebrates are complicated, as dispersal and recruitment can be influenced by, e.g., hydrodynamic mechanisms (including currents acting as vectors or barriers to dispersal), larval survival, pelagic larval duration, presence of suitable habitats for settlement, and presence of predators (Goode et al. 2020).

The degree of connectivity between populations is usually estimated using genetic markers (Vrijenhoek 2010). Connectivity estimation at an assemblage or community level may be attempted, as an approximation, by conducting a non-metric multidimensional similarity/dissimilarity analysis (e.g., a non-metric MDS). Such analyses for meiobenthic, microbenthic and megafaunal assemblages reported from the MAR vent fields sampled so far showed a fairly low between-field similarity, and thus limited connectivity, although—in the case of the meiobenthos—the two most closely located fields (Broken Spur and Lost City) were those in the Polish contract area. Interestingly, the vent field geographically closest to the Polish contract area (TAG) supported quite a dissimilar meiobenthic assemblage. In contrast, the Lost City macro- and megafaunal assemblages turned out to be most dissimilar from others. However, the patterns revealed by the MDS plots (e.g., in Fig. 1 in Chap. "Benthic Organisms") are by no means conclusive, as the available data base is scant and the benthic assemblages are under-sampled. Nevertheless, it may be expected that anthropogenic intervention due to PMS mining at the MAR (in individual contract areas and beyond them) will definitely put the ecosystem resilience, with all its attributes, to the test.

2 Hydrothermal Vent Fields in the Polish Exploration Contract Area on the Mid-Atlantic Ridge as Viewed from the Protection and Conservation Standpoint

In a recent assessment, commenced during the ISA-sponsored "Workshop on the Regional Environmental Management Plan for the Area of the Northern Mid-Atlantic Ridge" in Evora, Portugal (https://www.isa.org.jm/files/files/documents/Evora%20Workshop_3.pdf; 'the Evora workshop'; Gollner et al. 2021), of the environmental characteristics of the 11 hydrothermal fields known from the northern part of the Mid-Atlantic Ridge (MAR) subjected to ISA contracts for exploration (Gollner et al. 2021; cf. the Evora Workshop Report, website as above), the fields were categorised, based on the geological and biological attributes, in terms of their meeting the criteria which are thought to be important for environmental management, protection, and conservation. These criteria were: Uniqueness and Rarity; Functional Significance; Fragility; Life-History Traits of Component Species that Make Recovery Difficult; Structural Complexity; Biological Diversity; Biological Productivity; Naturalness; and Ecosystem Services. Each field was assigned a score with respect to each criterion (High, Medium, Low), the scores reflecting the extent to which the field met that particular criterion.

As repeatedly indicated in this volume, the Polish exploration contract area on the MAR encompasses two known hydrothermal vent fields: Lost City and Broken Spur.

The Lost City field (cf. Chaps. 3, 6), covering 10,000 km^2, located at a depth range of 740–850 m, is characterised by a relatively low-temperature (90 °C), high-pH (9–11), hydrogen- and methane rich vent fluid devoid of dissolved metals, and carbonate mineralogy (aragonite to calcite). The field has been intensively studied for its venting characteristics, geological set-up and microbiology, and less so in terms of the faunal communities it supports (but cf. Radziejewska et al. 2022). The Lost City was found to meet the criterion of Uniqueness and Rarity (score: High) in being distinct as an unusual serpentinite system, hosting unique microbial communities that may represent a "rare analogue for early Earth ecosystems" (e.g., Brazelton et al. 2010). The field scored High on the criterion Functional Significance, as it is exceptionally important for the survival of sulphur-cycling and methane-oxidising microorganisms adapted to high pH and methane-rich conditions; the vent-endemic taxa, including *Bathymodiolus azoricus*, are extremely rare, but the primary production of the field may be crucial for their reproduction. The Lost City scored High on the criterion Fragility, as it is thought that it may be susceptible to degradation by human activities (considering its uniqueness and the hosted microorganisms regarded as a rare biosphere). The Lost City scoring on the criterion Life-History Traits of Component Species that Make Recovery Difficult was also High, in view of the poor knowledge on recruitment dynamics and the fact that the field hosts species without larval dispersal (e.g., morphologically distinct nematodes). The Lost City scoring on the criterion Structural Complexity was High on account of within- and between-structure differences in mineralogy, carbonate age, fluid flux, chemistry, and associated microbiota and the presence of the engineer species, the vent-endemic *Bathymodiolus azoricus*. The Lost City scored High on the criterion Biological Diversity, primarily on account of the high diversity of microorganisms, and the diversity of invertebrates inhabiting the flanges and spires. In terms of the criterion Biological Productivity, the field scored High because of the abundant chemical energy for the synthesis of organic compounds. The scoring for the criterion Naturalness was High, as the Lost City is considered to be, at present, relatively undisturbed. Finally, the Lost City scoring on the criterion Ecosystem Services was also High, as the field was considered to offer—like all active vent fields—provisioning, regulating, and supporting services, the latter including, *inter alia*, inspiration for art, and serving as a centre for knowledge generation.

The Broken Spur field, with its ~5000 m^2 seafloor massive sulphide area, lies at a depth range of 3970–3110 m and is a 'typical' hydrothermal vent field in a basaltic hosting rock. It features seven known tall active high-temperature (360-364 °C) black smokers as well as diffuse-vent platforms and inactive weathered sulphide mounds. The venting fluid is enriched in lithium, and poor in dissolved manganese and strontium; the field contains appreciable (compared to other MAR fields) amounts of gold and silver. In terms of the criterion Uniqueness and Rarity, the Broken Spur was found to score High due to its hosting vent-endemic taxa, including the shrimps *Rimicaris ecoxulata*, *Alvinocaris markensis*, *Mirocaris fortunata*, *Keldyshicaris vavilovi*, mussels *Bathymodiolus azoricus*, *B. puteoserpentis* and their hybrid (O'Mullan et al. 2001; Breusing et al. 2016), crabs *Segonzacia mesatlantica*, galatheid squat lobsters *Munidopsis* sp., gastropods *Phymorhynchus moskalevi*, and ophiuroids *Ophioctenella acies*. The field scored High on the criterion Functional Significance as it is a discrete feeding area essential for the survival, reproduction and recovery of vent-endemic species listed above as well as two species of deep-sea fish *Haptenchelys texis* (Synaphobranchidae) and *Pachycara thermophilum* (Zoarcidae) (Parin 1995); it is a zone of hybridisation between the two *Bathymodiolus* mussel species (O'Mullan et al. 2001) and the source of copepodid stages of the vent-endemic dirivultid copepods. The field scored High on the criterion Fragility, the arguments being that it is a fragile, geographically isolated and discrete ecosystem, hosting populations of species documented as adversely affected by, e.g., illumination. The Broken Spur scoring on the criterion Life-History Traits of Component Species that Make Recovery Difficult was High, although the recovery dynamics of vent communities are unknown and many best-known species (e.g. *Bathymodiolus* mussels) have larval dispersal, but this trait is absent in many others. The field scored High on the criterion Structural Complexity, as it features several types of hydrothermal structures with different morphologies, chemical, and thermal conditions as well as the *Bathymodiolus* mussels as engineer species. The Broken Spur scoring on the criterion Biological Diversity was High on account of the field hosting vent-endemic invertebrate species and two rare species of deep-sea fishes (see above). The Broken Spur scored High on the criterion Biological Productivity: its hydrothermal venting supports high *in situ* primary production which in turn supports biomass-rich communities. The Broken Spur is relatively undisturbed at present, so it scored High on the criterion Naturalness. Finally, the Broken Spur scored High on the criterion Ecosystem Services, as it has been, and continues to be, an important research site yielding new biological discoveries, for instance hybrid mussels and new species (e.g., the shrimp *Keldyshicaris vavilovi*; Lunina and Vereshchaka 2010).

To sum up, both hydrothermal vent fields within the Polish exploration contract area on the MAR have been found to meet all the criteria of vulnerability and the need of protection.

3 Protection of Geological Formations

Hydrothermal vents are home to many remarkable organisms not found anywhere else in the world. Hence, discussion on the protection of these areas as valuable habitats has been carried out for years by many organizations. Extraction of massive sulphides from hydrothermal fields will physically destroy the chimneys, their fauna, and leave a flatter topography covered by compressed sediment. It is difficult to predict to what extent it will be possible for the habitats to recover and be re-colonized (Miller et al. 2018).

In order to reduce the negative human impacts on the deep sea, particularly valuable sites are defined as Marine Protected Areas (MPAs), marine reserves and no-take zones. Most of these areas are concentrated in exclusive economic zones (EEZs) or on respective extended continental shelves. An example is the establishment, in 2016, of the Azores Hydrothermal Vent MPA in the area under Portugal's national jurisdiction in the Atlantic Ocean (Ramirez-Llodra et al. 2011). The OSPAR Commission also recognized hydrothermal vents as valuable sites in need of protection, and placed them on the List of Threatened and/or Declining Species and Habitats (OSPAR 2008). Hydrothermal vents are also included within the category "reef" and are protected as one of the habitats within the EU Natura 2000 Network (European Commission DG Environment 2007).

In areas beyond national jurisdiction, the major role in the protection of the marine environment is to be played by the International Seabed Authority (ISA). Within its remit, the ISA is committed to ensuring the effective protection of the marine environment against harmful effects associated with mining activities on the international seabed, as stipulated by Article 145 of UNCLOS. As an endeavour to fulfil this commitment, Regional Environmental Management Plans (REMPs) are being developed, and Areas of Particular Environmental Interest (APEIs) have been designated in the Pacific's Clarion-Clipperton nodule field (Wedding et al. 2013). APEIs are to be excluded from future exploitation of mineral resources. On account of the uniqueness of active hydrothermal systems on the Mid-Atlantic Ridge, their legal protection has been deemed desirable (Freestone et al. 2016; Johnson 2019). However, a REMP for the Mid-Atlantic Ridge is currently in a preparatory phase (International Seabed Authority 2019).

The Polish contract area on the Mid-Atlantic Ridge encompasses two known (Beaulieu et al. 2013) hydrothermal fields: Broken Spur and Lost City. The latter is particularly important from the standpoint of protection of geological formations.

The Lost City is a hydrothermal field characterized by the occurrence of unique carbonate structures (Fig. 1). Its discovery in 2000 caused a sensation, and since then it has been intensively researched (Kelley et al. 2001, 2005; Seyfried et al. 2015; Lang and Brazelton 2020) (cf. Chap. 3, 6). The chimneys are 30 to 60 m tall and are composed of variable mixtures of calcite ($CaCO_3$), aragonite ($CaCO_3$), and brucite ($Mg(OH)_2$) (Dubinina et al. 2007; Kelley et al. 2001). They are located at a water depth of 750–900 m (Kelley et al. 2001, 2005). The fluid composition in Lost City hydrothermal field is derived from the reaction between seawater and uplifted mantle peridotite. This reaction produces a low temperature fluid (<40°–116 °C) with a high pH ranging from 9 to 11 (Kelley et al. 2001, 2005; Lang and Brazelton 2020). Lost City does not have any mineral resources of interest, so it probably will never be mined. However, adjacent vents may be mined. Plumes of sediment in the water column and possible chemical contamination due to, e.g., equipment failure may have an unpredictable effect on the functioning of hydrothermal vent communities (Johnson 2019).

Scientists, non-governmental organizations, and the international public opinion are concerned by the perceived threats to the ecosystems of hydrothermal vents. For instance, concern about the contracts for exploration signed by the ISA in the Mid-Atlantic Ridge has been expressed by Greenpeace (Casson et al. 2019). They fear

Fig. 1 An assemblage of carbonaceous 'chimneys' at the Lost City hydrothermal field (750–900 m depth), as photographed from the ROV Victor 6000. Source: IFREMER, https://image.ifremer.fr/data/00569/68094/

the destruction of the unique and still poorly understood ecosystems existing around hydrothermal vents. Of a particular concern for Greenpeace and other organisation is the Lost City, recognised as an Ecologically and Biologically Significant Area (EBSA) under the Convention on Biological Diversity (CBD) and still covered by the exploration contract signed between the ISA and Poland (Casson et al. 2019). Due to its outstanding universal value, the Lost City has been also recognised by the UNESCO as a potential World Heritage Site in the high seas (Freestone et al. 2016). On account of the unusual nature of the structures such as hydrothermal chimneys, every effort should be made to preserve these valuable natural geological formations.

References

Baldrighi E, Zeppilli D, Crespin R, Chauvaud P, Pradillon F, Sarrazin J (2018) Colonization of synthetic sponges at the deep-sea Lucky Strike hydrothermal vent field (Mid-Atlantic Ridge): a first insight. Mar Biodivers 48:89–103

Beaulieu SE, Baker ET, German CR, Maffei A (2013) An authoritative global database for active submarine hydrothermal vent fields. Geochem Geophys Geosyst 14:4892–4905. https://doi.org/10.1002/2013GC004998

Brazelton WJ, Ludwig KA, Sogin ML, Andreishcheva EN, Kelley DS, Shen C-C, Edwards RL, Baross JA (2010) Archaea and bacteria with surprising microdiversity show shifts in dominance over 1,000-year time scales in hydrothermal chimneys. Proc Natl Acad Sci 107:1612–1617. https://doi.org/10.1073/pnas.0905369107

Breusing C, Biastoch A, Drews A, Metaxas A, Jollivet D, Vrijenhoek RC, Bayer T, Melzner F, Sayavedra L, Petersen JM, Dubilier N, Schilhabel MB, Rosenstiel P, Reusch TBH (2016) Biophysical and population genetic models predict the presence of "phantom" stepping stones connecting Mid-Atlantic Ridge vent ecosystems. Curr Biol 26:2257–2267

Casson L, Losada S, Tsenikli S, Santillo D, Currie D, Sharma G, McCallum W (2019) In deep water: the emerging threat of deep sea mining. Greenpeace International. https://issuu.com/greenpeaceinternational/docs/in-deep-water-2019

Clark M, Smith S (2013) Environmental management considerations. In: Baker E, Beaudoin Y (eds) Deep sea minerals: sea-floor massive sulphides, a physical, biological, environmental, and technical review, vol 1A. Secretariat of the Pacific Community, pp 27–42

Cuvelier D, Beesau J, Ivanenko VN, Zeppilli D, Sarradin PM, Sarrazin J (2014) First insights into macro- and meiofaunal colonisation patterns on paired wood/slate substrata at Atlantic deep-sea hydrothermal vents. Deep Sea Res Part I Oceanogr Res Pap 87:70–81

Dubinina EO, Chernyshev IV, Bortnikov NS, Lein AY, Sagalevich AM, Gol'tsman YV, Bairova ED, Mokhov AV (2007) Isotopic–geochemical characteristics of the Lost City hydrothermal field. Geochem Int 45:1131–1143. https://doi.org/10.1134/S0016702907110067

European Commission DG Environment (2007) Natura 2000. Interpretation manual of European Union habitats. EUR, p 27

Freestone D, Laffoley D, Douvere F, Badman T (2016) World heritage in the high seas: an idea whose time has come. IUCN. https://unesdoc.unesco.org/ark:/48223/pf0000245467

Girard F, Lacharité M, Metaxas A (2016) Colonization of benthic invertebrates in a submarine canyon in the NW Atlantic. Mar Ecol Prog Ser 544:53–64

Goode SL, Rowden AA, Bowden DA, Clark MR (2020) Resilience of seamount benthic communities to trawling disturbance. Mar Environ Res 161:105086, ISSN 0141-1136. https://doi.org/10.1016/j.marenvres.2020.105086

Gollner S, Kaiser S, Menzel L, Jones DOB, Brown A, Mestre NC, Van Oevelen D, Menot L, Colaco A, Canals M, Cuvelier D, Durden JM, Gebruk A, Egho GA, Haeckel M, Marcon Y,

Mevenkamp L, Morato T, Pham CK, Purser A, Sanchez-Vidal A, Vanreusel A, Vink A, Martinez P, van Oevelen D, Menot L, Colaço A, Canals M, Cuvelier D, Durden JM, Gebruk A, Egho GA, Haeckel M, Marcon Y, Mevenkamp L, Morato T, Pham CK, Purser A (2017) Resilience of benthic deep-sea fauna to mining activities. Mar Environ Res 129:76–101

Gollner S, Colaço A, Gebruk A, Halpin PH, Higgs N, Menini E, Mestre NC, Quian P-Q, Sarrazin J, Szafranski K, Van Dover CL (2021) Application of scientific criteria for identifying hydrothermal ecosystems in need of protection. Mar Policy 132:104641. (Annex II, Evora Workshop Report; https://www.isa.org.jm/files/files/documents/Evora%20Workshop_3.pdf)

International Seabed Authority (2019, November 25–29) Report of the Workshop on the regional environmental management plan for the area of the Northern Mid-Atlantic Ridge, Évora, Portugal. https://www.isa.org.jm/files/2020-01/Evora%20Workshop.pdf

Johnson DE (2019) Protecting the Lost City hydrothermal vent system: all is not lost, or is it? Marine Policy 107:103593. https://doi.org/10.1016/j.marpol.2019.103593

Kelley DS, Karson JA, Blackamn DK, Früh-Green GL, Butterfield DA, Lilley MD, Olson EJ, Schrenk MO, Roe KK, Lebon GT, Rivizzigno P, the AT3-60 Shipboard Party (2001) An off-axis hydrothermal vent field near the Mid-Atlantic Ridge at 30°N. Nature 412:145–149. https://doi.org/10.1038/35084000

Kelley DS, Karson JA, Früh-Green GL, Yoerger DR, Shank TM, Butterfield DA, Hayes JM, Schrenk MO, Olson EJ, Proskurowski G, Jakuba M, Bradley A, Larson B, Ludwig K, Glickson D, Buckman K, Bradley AS, Brazelton WJ, Roe K, Elend MJ, Delacour A, Bernasconi SM, Lilley MD, Baross JA, Summons RE, Sylva SP (2005) A serpentinite-hosted ecosystem: the Lost City hydrothermal field. Science 307:1428–1434. https://doi.org/10.1126/science.1102556

Lacharité M, Metaxas A (2013) Early life history of deep-water gorgonian corals may limit their abundance. PLoS One 8:e65394

Lang SQ, Brazelton WJ (2020) Habitability of the marine serpentinite subsurface: a case study of the Lost City hydrothermal field. Philos Trans R Soc A 378:20180429. https://doi.org/10.1098/rsta.2018.0429

Levin LA, Wei C-L, Dunn DC, Amon DJ, Ashford OS, Cheung WWL, Colaço A, Dominguez-Carrió C, Escobar EG, Harden-Davies HR, Drazen JC, Ismail K, Jones DOB, Johnson DE, Le JT, Lejzerowicz F, Mitarai S, Morato T, Mulsow S, Snelgrove PVR, Sweetman AK, Yasuhara M (2020) Climate change considerations are fundamental to management of deep-sea resource extraction. Glob Chang Biol 26:4664–4678

Lotze HK, Coll M, Magera AM, Ward-Paige C, Airoldi L (2011) Recovery of marine animal populations and ecosystems. Trends Ecol Evol 26:505–605

Lunina AA, Vereshchaka AL (2010) A new vent shrimp (Crustacea: Decapoda: Alvinocarididae) from the Mid-Atlantic Ridge. Zootaxa 2372:69–74

Miller KA, Thompson KF, Johnston P, Santillo D (2018) An overview of seabed mining including the current state of development, environmental impacts, and knowledge gaps. Front Mar Sci 4:418. https://doi.org/10.3389/fmars.2017.00418

Mullineaux LS, Butman CA (1990) Recruitment of encrusting benthic invertebrates in boundary-layer flows: a deep-water experiment on cross seamount. Limnol Oceanogr 35:409–423

O'Mullan GD, Maas PA, Lutz RA, Vrijenhoek RC (2001) A hybrid zone between hydrothermal vent mussels (Bivalvia: Mytilidae) from the Mid-Atlantic Ridge. Mol Ecol 10:2819–2831

OSPAR Commission, OSPAR (2008) List of threatened and/or declining species and habitats, Reference Number: 2008-6. https://www.ospar.org/work-areas/bdc/species-habitats/list-of-threatened-declining-species-habitats. Accessed 20 October 2020

Parin NV (1995) Deep-sea fishes, *Haptenchelys texis* (Synaphobranchidae) and *Pachycara thermophilum* (Zoarcidae) from the Broken Spur hydrothermal vent field (Mid-Atlantic Ridge). J Ichthyol 35:328–332

Radziejewska T, Błażewicz M, Włodarska-Kowalczuk M, Jóźwiak P, Pabis K, Węsławski JM (2022) Benthic biology in the Polish exploration contract area of the Mid-Atlantic Ridge: the knowns and the unknowns. A review. Front Mar Sci 9:898828

Ramirez-Llodra E, Tyler PA, Baker MC, Bergstad OA, Clark MR, Escobar E, Menot L, Rowden AA, Smith CR, Van Dover CL (2011) Man and the last great wilderness: human impact on the deep sea. PLoS One 6:e22588. https://doi.org/10.1371/journal.pone.0022588

Seyfried WE, Pester NJ, Tutolo BM, Ding K (2015) The Lost City hydrothermal system: constraints imposed by vent fluid chemistry and reaction path models on subseafloor heat and mass transfer processes. Geochim Cosmochim Acta 163:59–79. https://doi.org/10.1016/j.gca.2015.04.040

Van Dover CL (2011) Mining seafloor massive sulphides and biodiversity: what is at risk? J Mar Sci 68(2):341–348

Van Dover CL (2014) Impacts of anthropogenic disturbances at deep-sea hydrothermal vent ecosystems: a review. Mar Environ Res 102:59–72

Van Dover CL (2019) Inactive sulfide ecocystems in the deep sea: a review. Front Mar Sci 6:461

Van Dover CL, Colaço A, Collins PC, Croot P, Metaxas A, Murton BJ, Swaddling A, Boschen-Rose RE, Carlsson J, Cuyvers L, Fukushima T, Gartman A, Kennedy R, Krieten C, Mestre NC, Molodtsova T, Myhrvold A, Pelleter E, Popoola SO, Qiant P-Y, Sarrazin J, Sharma R, Suh YJ, Sylvan JB, Tao C, Tomczak M, Vermil J (2020) Research is needed to inform environmental management of hydrothermally inactive and extinct polymetallic sulfide (PMS) deposits. Mar Policy 121:104183

Vrijenhoek RC (2010) Genetic diversity and connectivity of deep-sea hydrothermal vent metapopulations. Mol Ecol 19:4391–4411

Wedding LM, Friedlander AM, Kittinger JN, Watling L, Gaines SD, Bennett M, Hardy SM, Smith CR (2013) From principles to practice: a spatial approach to systematic conservation planning in the deep sea. Proc R Soc B 280:20131684

Yearsley JM, Salmanidou DM, Carlsson J, Burns D, Van Dover CL (2020) Biophysical models of persistent connectivity and barriers on the northern Mid-Atlantic Ridge. Deep Sea Res Part II Top Stud Oceanogr 180:104819. https://doi.org/10.1016/j.dsr2.2020.104819

Zeppilli D, Vanreusel A, Pradillon F, Fuchs S, Mandon P, James T, Sarrazin J (2015) Rapid colonisation by nematodes on organic and inorganic substrata deployed at the deep-sea Lucky Strike hydrothermal vent field (mid-Atlantic ridge). Mar Biodivers 45:489–504

The Legal and Regulatory Framework for Protection and Conservation of the Deep-Sea Floor, with a Particular Reference to Mid-Ocean Ridges: A Summary

Teresa Radziejewska, Agata Kozłowska-Roman, and Agata Szyduk

Activities aimed at protection and conservation of the deep-sea floor are at present governed by legal instruments which form a multi-tier system (cf. Jaeckel et al. 2020). On the most general level, the relevant issues are treated by the United Nations Convention on the Law of the Sea of 1982 (UNCLOS, www.un.org/depts/los/convention_agreements/texts/unclos/UNCLOS-TOC.htm), and in particular by Part XI of the Convention and the Agreement on the Implementation of Part XI of 16 November 1994 (the date when UNCLOS entered into force; Van Dover et al. 2011) (www.un.org/Depts/los/convention_agreements/texts/unclos/closindxAgree.htm).

Auxiliary and complementary to UNCLOS are resolutions and documents issued by the UN General Assembly. Of a particular importance here is the 2015 Agenda for Sustainable Development 2030 which lists 17 Sustainable Development Goals (SDGs). In the context of deep-sea environment protection, particularly relevant is SDG 14 ("Conserve and sustainably use the oceans, seas and marine resources for sustainable development").

Protection and conservation of marine environment, its biodiversity in particular, are dealt with in a number of other international instruments, auxiliary to the UNCLOS, including the Stockholm Declaration (https://legal.un.org/avl/ha/dunche/dunche.html) signed following the 1972 United Nations Conference on the Human Environment (the so-called Stockholm Conference; www.un.org/ga/search/view_doc.asp?symbol=A/CONF.48/14/REV.1) which provided the basis for sustainable development as a core principle for the management of human impact on the

T. Radziejewska (✉)
Institute of Marine and Environmental Sciences, University of Szczecin, Szczecin, Poland
e-mail: teresa.radziejewska@usz.edu.pl

A. Kozłowska-Roman · A. Szyduk
Polish Geological Institute – National Research Institute, Warsaw, Poland
e-mail: akozl@pgi.gov.pl

T. Radziejewska et al. (eds.), *The Natural Environment of the Mid-Atlantic Ridge*, GeoPlanet: Earth and Planetary Sciences,
https://doi.org/10.1007/978-3-031-51865-2_9

environment. Subsequent declarations and instruments, including the Rio Declaration on Environment and Development of 1992 (https://www.un.org/en/development/desa/population/migration/generalassembly/docs/globalcompact/A_CONF.151_26_Vol.I_Declaration.pdf) and the Convention on Biological Diversity (CBD) of 1993 (www.cbd.int) have upheld the principle and developed it further. Although recognising that ecosystems, species and genes must be used for the benefit of mankind, this use should not result in a long-term decline in biodiversity and environmental damage, hence the CBD's objective is conservation of all biodiversity. The CBD supports the Precautionary Principle (McIntyre and Mosedale 1997) according to which the burden of proof should be shifted to those who intend to undertake or continue an activity that poses a threat of serious or irreversible damage. According to the CBD, the ecosystem approach (e.g., Rice et al. 2005) should be a strategy of preference for integrated and adaptive management to promote conservation and sustainable use of the environment. Moreover, the CBD promotes the use of area-based management tools (envisaged also for mid-oceanic ridge areas to be explored and developed; Van Dover et al. 2011). Consequently, the CBD lists 7 scientific criteria with which to identify, on a global scale, the so-called Ecologically and Biologically Significant Areas (EBSAs). Other international initiatives associated with the area-based management tools includes a call for the establishment of representative networks of marine protected areas (MPAs) consistent with international law and based on scientific information. The list of relevant resolution of the UN General Assembly (UNGA) includes UNGA 61/105 (2006) and UNGA 64/72 (2009) aimed at establishing the so-called Vulnerable Marine Ecosystems (VMEs) (a concept also applied to the mid-oceanic ridges, notably the Mid-Atlantic Ridge; Van Dover et al. 2012; Dunn et al. 2018).

Of relevance is the recent Agreement under the UN Convention on the Law of the Sea (UNCLOS) on the Conservation and Sustainable Use of Marine Biological Diversity of Areas Beyond National Jurisdiction (BBNJ) (www.un.org/bbnj/).

The UNCLOS, UNGA resolutions, and the CBD sparked several regional arrangements aimed at implementation of measures for the protection of the marine environment in areas beyond national jurisdiction. Most relevant here is the OSPAR Convention for the Protection of the Marine Environment of the North east Atlantic (www.ospar.org/convention) as well as Regional Fisheries Management Organizations (RFMOs). The latter are obliged to conserve all species associated with and dependent upon the fisheries they seek to regulate (Bell et al. 2019) and acting according to, *inter alia*, the FAO Code of Conduct for Responsible Fisheries (http://www.fao.org/fishery/code/en) and the UN Fish Stocks Agreement (http://www.fao.org/iuu-fishing/international-framework/un-fish-stocks-agreement/en/); some RFMOs have already enacted fisheries closures in areas beyond national jurisdiction (VMEs).

According to UNCLOS, marine seafloor areas beyond the areas of national jurisdictions (termed "the Area") are regarded as "the common heritage of mankind"; their resources and overall management, including protection and conservation, form the remit of the International Seabed Authority (ISA) as the institution through which States Parties to UNCLOS are to organize and control prospecting, exploration and exploitation of mineral resources in the Area. Such activities may not be carried out without a contract with the ISA in accordance with its rules, regulations and procedures. UNCLOS Article 145 obliges ISA to provide an efficient protection of the marine environment from any harm which may be caused by activities in the Area. In particular, the ISA is obliged to take necessary measures with respect to activities in the Area "to ensure effective protection for the marine environment from harmful effects which may arise from such activities. To this end the Authority shall adopt appropriate rules, regulations and procedures for inter alia: (a) the prevention, reduction and control of pollution and other hazards to the marine environment, including the coastline, and of interference with the ecological balance of the marine environment, particular attention being paid to the need for protection from harmful effects of such activities as drilling, dredging, excavation, disposal of waste, construction and operation or maintenance of installations, pipelines and other devices related to such activities; (b) the protection and conservation of the natural resources of the Area and the prevention of damage to the flora and fauna of the marine environment" (UNCLOS, Article 145).

Consequently, the second level of the legal system are the regulations issued by the ISA and its organ called the Legal and Technical Commission (LTC). Of relevance here are the ISA Regulations on prospecting and exploration for polymetallic sulphides in the Area (adopted 7 May 2010) (www.isa.org.jm). These regulations contain requirements related to environmental protection, including obligations of contractors to collect and report environmental baseline data, carry out environmental monitoring programmes supervised by the ISA, and to establish environmental baselines against which impacts from anticipated mining activities can be assessed. The LTC issues recommendations to guide contractors in performing these obligations (the most recent is ISBA/25/LTC/6/Rev.1/Corr.1 of 11 June 2020) and regularly reviews reports and data submitted by the contractors, with the assistance of scientists and other stakeholders. At present, the ISA is in the process of working out regulations on exploitation activities (the so-called Mining Code) such that they would not cause excessive harm to the marine environment (Levin et al. 2016). This is a formidable task given that commercial deep-sea mineral mining has not begun yet.

Based on UNCLOS Part XI and the 1994 Agreement, the ISA has developed detailed rules, regulations and recommendations for the contractors regarding the latter's obligation to carry out the assessment of potential environmental impacts (on the water column and the seafloor, including biodiversity) of exploration for marine minerals in the Area; those regulations (the most recent are contained in ISBA/26/LTC/CRP.12 of 4 March 2020) describe in detail the type of activities which require preparation of the formal Environmental Impact Assessment (EIA) as well as its format and content, and provide recommendations on baseline research, monitoring, and reporting.

In parallel to the work on the Mining Code, the ISA and its stakeholders are involved in developing, for each resource type, a Regional Environmental Management Plan (REMP) to be used by the ISA and its contractors as a tool with which to resolve the sustainable use of the mineral resources with the need to protect and conserve the marine environment, and as way to attain the SDG 14. The first REMP, approved by the ISA Council in 2012 (ISBA/18/C/22), concerned the Clarion-Clipperton Fracture Zone nodule field and established nine Areas of Particular Environmental Interest (APEI) to be excluded from future exploitation (Wedding et al. 2013). At present, work is in progress on developing REMPs for mid-oceanic ridges, including the Mid-Atlantic Ridge (MAR) (e.g., www.isa.org.jm/files/files/documents/paris_meeting_report_for_webposting.pdf).

On the third level of the legal system discussed are national regulations governing activities of contractors based in (and/or endorsed by) individual countries. As yet, there are no Polish regulations directly applicable to exploration and exploitation of deep-sea resources which would govern exploration of the Polish contract area on the MAR. Exploration activities to be carried out there are being prepared based on the Programme for Geological Exploration of the Oceans (PROGEO) adopted in 2017 by the Council of Ministers of the Republic of Poland (Anonymus 2017).

In addition to the three-level system outlined above, codes of conduct related to environmental protection in the deep sea have been prepared by some of the principal stakeholder groups' (scientists and the mining industry) organisations. The most relevant of these include InterRidge and the International Marine Minerals Society (IMMS). InterRidge is a non-profit international scientific organisation with a mission to drive oceanic ridge research forward in a way that is cost-effective, cooperative and responsible. The InterRidge Statement of Commitment to Responsible Research Practices (http://www.interridge.org/IRStatement) at deep-sea hydrothermal vents affirms the commitment of the InterRidge community to responsible research (Glowka 2003). The IMMS is a professional non-profit organisation the members of which are interested in marine minerals as resources with which to meet world demand for strategic minerals. The IMMS has drafted a Code for Environmental Management of Marine Mining (http://www.immsoc.org/IMMS_code.htm) which outlines environmental principles for the marine mining industry and operating guidelines that can be used as a basis for environmental management plans (Jones and Morgan 2003; Verlaan 2010). More general issues associated with corporate responsibility in the protection of the natural environment have been recently dealt with by Morgera (2020) and may be applicable to entities (contractors) carrying exploration of the deep seafloor, including on mid-oceanic ridges.

References

Anonymus (2017) Program Rozpoznania Geologicznego Oceanów. Monitor Polski 2:77

Bell JB, Guijarro-Garcia E, Kenny A (2019) Demersal fishing in areas beyond National Jurisdiction: a comparative analysis of regional fisheries management organisations. Front Mar Sci 6:596. https://doi.org/10.3389/fmars.2019.00596

Dunn DC, Van Dover CL, Etter RJ, Smith CR, Levin LA, Morato T, Colaço A, Dale AC, Gebruk AV, Gjerde KM, Halpin PN, Howell KL, Johnson D, Perez JAA, Ribeiro MC, Stuckas H, Weaver P (2018) A strategy for the conservation of biodiversity on mid-ocean ridges from deep-sea mining. Sci Adv 4:eaar4313

Glowka L (2003) Putting marine scientific research on a sustainable footing at hydrothermal vents. Mar Policy 27:303–312

Jaeckel A, Gjerde K, Currie D (2020) The legal framework for resource management in the deep sea. In: Baker M, Ramirez-Llodra E, Tyler P (eds) Natural capital and exploitation of the deep ocean. Oxford University Press, pp 53–69

Jones AT, Morgan CL (2003) Code of practice for ocean mining: an international effort to develop a code for environmental management of marine mining. Mar Geosci Geotechnol 21:105–114

Levin LA, Mengerink K, Gjerde KM, Rowden AA, Van Dover CL, Clark MR, Ramirez-Llodra A, Currie B, Smith CR, Satoi KN, Galloi N, Sweetman AK, Lily H, Armstrong CW, Brider J (2016) Defining "serious harm" to the marine environment in the context of deep-seabed mining. Mar Policy 74:245–259

McIntyre O, Mosedale T (1997) The precautionary principle as a norm of customary international law. J Environ Law 9:221–241

Morgera E (2020) Corporate environmental accountability in international law, 2nd edn. Oxford University Press

Rice J, Trujillo V, Jennings S, Hylland K, Hagstrom O, Astudillo A, Jensen JN (2005) Guidance on the application of the ecosystem approach to Management of Human Activities in the European marine environment. ICES Cooperative Research Report, No. 273, ICES, Copenhagen

Van Dover CL, Smith CR, Ardron J, Arnaud S, Beaudoin Y, Bezaury J, Boland G, Billett D, Carr M, Cherkashov G, Cook A, DeLeo F, Dunn D, Fisher CR, Godet L, Gjerde K, Halpin P, Levin L, Lodge M, Menot L, Miller K, Milton D, Naudts L, Nugent C, Pendleton L, Plouviez S, Rowden A, Santos R, Shank T, Smith S, Tao C, Tawake A, Thurnherr A, Treude T (2011) Environmental management of deep-sea chemosynthetic ecosystems: justification of and considerations for a spatially-based approach. ISA Technical Study, No. 9. International Seabed Authority, Kingston, Jamaica

Van Dover CL, Smith CR, Ardron J, Dunn D, Gjerde K, Levin L, Smith S (2012) Designating networks of chemosynthetic ecosystem reserves in the deep sea. Mar Policy 36:378–381

Verlaan P (2010) The international marine mineral society's code for environmental management of marine mining. Offshore Technology Conference. https://doi.org/10.4043/20914-MS

Wedding LM, Friedlander AM, Kittinger JN, Watling L, Gaines SD, Bennett M, Hardy SM, Smith CR (2013) From principles to practice: a spatial approach to systematic conservation planning in the deep sea. Proc R Soc B 280:20131684

Ecosystem Valuation and Monitoring

Jan Marcin Węsławski, Teresa Radziejewska, and Michał Tomczak

1 Valued Ecosystem Components: Risk Assessment

Environmental Impact Assessment (EIA) is by definition an unbiased expert judgement, presenting possible impacts of the planned activity on the natural world. In such an assessment, environmental experts stand behind the Nature to sound a warning against possible consequences of human intervention. Impacts may affect the structure (presence/absence of species, existence of habitats) and the function (energy flow, food web alteration, seasonality disruption, life cycles) of an ecosystem (cf. According to the definitions of the 2010 Millenium Assessment (https://www.millenniumassessment.org), the system of goods and services provided by each ecosystem is a subject of valuation (cf. Armstrong et al. 2012; Thurber et al. 2014; Folkersen et al. 2018), and EIA shall indicate possible alterations there. Another approach is to define the Valued Ecosystem Components (VEC) and focus on the likely impacts they may be exposed to. Table 1 presents an approach integrating both aspects. The three main types of valuation are: (1) socioeconomic valuation, with reference to the market value of ecosystem goods and services; (2) biological valuation, presenting the intristinc value of a natural ecosystem, without reference to its human use; and (3) socio-cultural valuaton that considers the

J. M. Węsławski (✉)
Institute of Oceanology PAN, Sopot, Poland
e-mail: weslaw@iopan.gda.pl

T. Radziejewska
Institute of Marine and Environmental Sciences, University of Szczecin, Szczecin, Poland
e-mail: teresa.radziejewska@usz.edu.pl

M. Tomczak
Polish Geological Institute – National Research Institute, Warszawa, Poland
e-mail: mtomc@pgi.gov.pl

T. Radziejewska et al. (eds.), *The Natural Environment of the Mid-Atlantic Ridge*, GeoPlanet: Earth and Planetary Sciences,
https://doi.org/10.1007/978-3-031-51865-2_10

Table 1 Valued ecosystem components (VEC) and ecosystem goods and services associated with the MAR ecosystem

Assessment component	Term	Type of value	Monitoring method	Resilience
VEC 1	Benthic Sea scape	Biological and socio-cultural	Remote video recording	Extremaly low
VEC 2	Rare and charismatic species	Biological and socio-cultural	Ship-based observations	Medium, species-dependent
VEC 3	Potentially commercial species	Socio-economic	Regular sampling and remote hydroacoustic survey	Medium, depending on species
VEC 4	Pelagic-benthic coupling (connectivity)	Socio-economic, biological and socio-cultural	Regular sampling	Probably high, considering the multitude of its components

non-market value of a spiritual, educational and emotional character. An important part of the EIA is the initial risk assessment, as some VECs might be very resilient, other—extremaly vulnerable; some are able to regenerate or restore, while others are almost unrepairable.

The Environmental Impact Assessment (EIA) is a process (Collins et al. 2013; Durden et al. 2018; Clark 2019) aimed at (i) describing the environment before an activity the impacts of which are to be assessed is commenced; (ii) assessing the effects of that activity (impacts); (iii) indicating follow-up activities and mitigation measures (Clark 2019). As shown by Clark (2019), the process consists of three basic components, each being a responsibility of a different actor/group of actors. The first component is the overall environmental strategy framework, developed by the regulator (International Seabed Authority, ISA), and including a relevant regional environmental management plan (REMP). The second component is embedded in the first, and is the remit of a contractor who is required to develop its internal environmental management plans and environmental management system, conduct an EIA and risk assessment, and carry out the follow-up activities (monitoring and mitigation; cf. Cuvelier et al. 2018). Finally, the third component involves an external assessment (Lallier and Maes 2016; Durden et al. 2018), to be conducted by state sponsors, stakeholders, and financing institutions.

It is thus evident that the fundamental part of the EIA process is the responsibility of the contractor, required—starting from the exploration phase—to provide a baseline assessment of the environment and to provide an evaluation of possible impacts arising from exploration activities. In fulfilling the obligations associated with the EIA process, the contractor is bound by the regulations, recommendations and guidelines issued by the International Seabed Authority (ISA). These regulations, binding for the contractor, are currently contained in three documents: ISBA/16/A/12/Rev.1 (2010), ISBA/26/LTC/CRP.12 (2020), and ISBA/25/LTC/6/Rev.1 (2020). The contractor is also mandated to conduct environmental monitoring, both as a way to collect baseline data and as an underpinning of the sound EIA.

2 Monitoring: Benthos

In the classical sense, monitoring is "a programme of recurring, systematic studies that reveal the state of the environment. The specific aspects of the environment to be studied are determined by environmental objectives and environmental legislation" (https: //ec.europa.eu/jrc/en/research-topic/environmental-monitoring). The goals of those observations of and studies on the object(s) being monitored are three-fold: (i) to follow changes in the object(s) monitored to provide information on their current status; (ii) given certain environmental objectives being set, to assess the progress made to achieve those objectives; and (iii) to help to detect new environmental issues. As it is evident from the above, the stress is on recurrence (in temporal and spatial scale) and systematicity (in terms of methods applied) of monitoring activities.

The monitoring of deep-sea areas, including those earmarked for exploration and future resource development (i.e., deep-sea mining), as carried out to date has not met any of those criteria. Instead, monitoring activities undertaken (and reported) by contractors so far involve collection of baseline data, extremely needed, but obtained in a rather haphazard manner in terms of temporal and spatial resolution. This is, however, understandable in view of the vastness of the area(s) to be surveyed as well as the financial and temporal constraints. Nevertheless, monitoring—whether at the exploration or exploitation stage—is a prerequisite of acquiring a reliable data sets on which to base reliable environmental impact assessments (Jones et al. 2019), including cumulative impacts (Le Bris and Levin 2020), and to disentangle natural variability from changes introduced by anthropogenic intervention.

This conviction is reflected in the International Seabed Authority (ISA) document ISBA/16/A/12/rev. 1 ("Contractors, sponsoring States and other interested States or entities shall cooperate with the Authority in the establishment and implementation of programmes for monitoring and evaluating the impacts of deep seabed mining on the marine environment.") as well as in the most recent ISA documents ISBA/26/LTC/CRP.12 and ISBA/25/LTC/6/Rev.1.

Since key impacts from the development of deep seafloor resources (including polymetallic sulphides) are expected to be incurred by the benthic communities (e.g., Van Dover 2014; Gollner et al. 2017; Danovaro et al. 2020), which is also reflected in the ISA document ISBA/25/LTC/6/Rev.1 ("On the basis of current proposed methodologies, the main impacts are expected to occur at the sea floor"), it is obvious that those communities should be monitored in the first place. This is reflected in recommendations for contractors regarding the assessment of environmental impacts of exploration (ISBA/25/LTC/6/Rev.1 2020) which list the following information specific to polymetallic sulphide exploration required from contractors:

"(a) Any modification of fluid discharge in hydrothermal settings and associated fauna (using photo documentation, temperature measurements and other metrics, as appropriate) …; (b) For active sulphide deposits, temperature-fauna relationships … (e.g. 5 to 10 discrete, video-documented temperature measures within each subhabitat); (c) The distribution, abundance, species structure and diversity of the dominant taxa in each subhabitat (active, inactive vent areas, non-vent habitats) …. This includes the assessment of specialist localized chemosynthetic communities relative to potential mining locations; (d) Meiofaunal and microbial community structure and biomass associated with the polymetallic sulphide deposits ... A statistically defensible number of samples should be taken from polymetallic sulphides, where possible. Species that live on the rock or in crevices and pits in the deposit should be identified, where possible; (e) Biological samples from active hydrothermal vent systems should only be collected using precision sampling by remotely operated vehicle/submersible technology according to subhabitat".

In addition, according to the documents mentioned, the contractors will be required to establish an environmental baseline against which the natural variability, including that imposed by climate change (cf. Le Bris and Levin 2020), and impacts produced by mining will be assessed. The contractors will be also obliged to monitor and evaluate mining impacts on the marine environment by surveying the distribution and composition of faunal communities. The contractors will have to classify hydrothermal vent areas according to whether they are active vent sites or inactive/extinct sites (cf. Van Dover 2019). From the biological standpoint, the baseline assessment should determine whether the proposed mining site supports active hydrothermal vents, inactive vents that may restart owing to mining activity, or extinct vents that will remain hydrothermally inactive even when disturbed by test-mining.

Monitoring of the benthos in the exploration/exploitation area should identify communities and their ecosystem functions, including their natural spatial and temporal variability, so as to evaluate potential effects of the activities on the benthic fauna. The benthic communities should be characterised in all the subhabitats that may be impacted; such surveys are also important for establishing preservation and impact reference zones necessary for environmental management of the seafloor areas by the ISA (Jones et al. 2019).

The sampling design applied in monitoring surveys should be well thought-out and robust, including stratified random sampling approaches, if variation of ecological conditions or habitats is known or expected. With respect to the macro- and meiobenthos, analyses should be appropriate and detailed enough to be able to establish a species accumulation curve (from which to estimate the number of samples required to result in an adequate assessment of taxonomic richness) (ISBA/25/LTC/6/Rev.1 2020).

The benthic categories the contractors will be required to monitor include the megafauna, macrofauna, meiobenthos, microeukaryotes (Foraminifera) as well as microorganisms (bacteria, archaea, fungi, viruses, microeukaryotes). In addition, environmental DNA (eDNA) will need to be analysed. Whenever possible, colour photographic documentation of organisms should be prepared (ISBA/25/LTC/6/Rev.1 2020).

Prior to test mining or testing of mining components, spatial variation in the composition of the biological community and levels of connectivity must be evaluated, as it is important—for the future impact assessments—to know the degree of isolation of populations occupying the mineral deposits that are to be removed and to infer whether a given population serves as a critical brood stock for other populations (cf. Gollner et al. 2017). Sampling should occur at three mineral deposit sites as well as at background communities in the contract area. In some cases (particularly with respect to the megafauna), video documentation and photographic transects may be the only way to develop a species-abundance matrix (ISBA/25/LTC/6/Rev.1 2020).

For monitoring of the benthos, it is also important to have knowledge on bioturbation (i.e., the mixing of sediment by benthic organisms) as a way to monitor the rates of sedimentary processes as well as to model and evaluate effects of mining activities on such processes. Bioturbation rates and depth will have to be measured (preferably by establishing excess Pb^{210} profiles in sediment cores), taking into account the variability in the sediment (ISBA/25/LTC/6/Rev.1 2020).

References

Armstrong CW, Foley NS, Tinch R, van den Hove S (2012) Services from the deep: steps towards valuation of deep sea. Ecosyst Serv 2:2–13

Clark M (2019) The development of environmental impact assessments for deep-sea mining. In: Sharma R (ed) Environmental issues of deep-sea mining. Springer, pp 447–470

Collins PC, Croot P, Carlsson J, Colaço A, Grehan A, Hyeong K, Kennedy R, Mohn C, Smith S, Yamamoto H, Rowden AA (2013) A primer for the environmental impact assessment of mining at seafloor massive sulfide deposits. Mar Policy 42:198–209

Cuvelier D, Gollner S, Jones DOB, Kaiser S, Arbizu PM, Menzel L, Mestre NC, Morato T, Pham C, Pradillon F, Purser A, Raschka U, Sarrazin J, Simon-Lledó E, Stewart IM, Stuckas H, Sweetman AK, Colaço A (2018) Potential mitigation and restoration actions in ecosystems impacted by seabed mining. Front Mar Sci 5:467. https://doi.org/10.3389/fmars.2018.00467

Danovaro R, Fanelli E, Aguzzi J, Billet D, Carugati L, Corinaldesi C, Dell'Anno A, Gjerde K, Jamieson AJ, Kark S, McClain C, Levin L, Levin N, Ramirez-Llodra E, Ruhl H, Smith CR, Snelgrove PVR, Thomsen L, Van Dover CL, Yasuhara M (2020) Ecological variables for developing a global deep-ocean monitoring and conservation strategy. Nat Ecol Evol 4:181–192

Durden JM, Lallier LE, Murphy K, Jaeckel A, Gjerde K, Jones DOB (2018) Environmental impact assessment process for deep-sea mining in 'the area'. Mar Policy 87:194–202

Folkersen MV, Fleming CM, Hasan S (2018) The economic value of the deep sea: a systematic review and meta-analysis. Mar Policy 94:71–80

Gollner S, Kaiser S, Menzel L, Jones DOB, Brown A, Mestre N, van Oevelen D, Menot L, Colaço A, Canals M, Cuvelier D, Durden JN, Gebruk A, Egho GA, Haeckel M, Marcon Y, Mevenkamp L, Morato T, Pham CK, Purser A, Sanchez-Vidal A, Vanreusel A, Vink A, Martinez Arbizu P (2017) Resilience of benthic deep-sea fauna to mining activities. Mar Environ Res 129:76–101

ISBA/16/A/12/Rev.1 (2010, 26 April–7 May) Decision of the assembly of the international seabed authority relating to the regulations on prospecting and exploration for polymetallic sulphides in the area. ISBA, Kingston
ISBA/25/LTC/6/Rev.1 (2020, Mar 30) Recommendations for the guidance of contractors for the assessment of the possible environmental impacts arising from exploration for marine minerals in the area. ISBA, Kingston
ISBA/26/LTC/CRP.12 (2020, Mar 4) Standard for ISA environmental impact assessment process. ISBA, Kingston
Jones DOB, Durden JM, Murphy K, Gjerde KM, Gebicka A, Colaco A, Morato T, Cuvelier D, Billet DSM (2019) Existing environmental management approaches relevant to deep-sea mining. Mar Policy 103:172–181
Lallier LE, Maes F (2016) Environmental impact assessment procedure for deep seabed mining in the area: independent expert review and public participation. Mar Policy 70:212–219
Le Bris N, Levin LA (2020) Climate change cumulative impacts on deep-sea ecosystems. In: Baker M, Ramirez-Llodra E, Tyler P (eds) Natural capital and exploitation of the deep ocean. Oxford University Press, pp 161–182
Thurber AR, Sweetman AK, Narayanaswamy BE, Jones DOB, Ingels J, Hansman RL (2014) Ecosystem function and services provided by the deep sea. Biogeosciences 11:3941–3963
Van Dover CL (2014) Impacts of anthropogenic disturbances at deep-sea hydrothermal vent ecosystems: a review. Mar Environ Res 102:59–72
Van Dover CL (2019) Inactive sulfide ecosystems in the deep sea: a review. Front Mar Sci 6:461

The manufacturer's authorised representative in the EU is Springer Nature Customer Service Centre GmbH, Europaplatz 3, 69115 Heidelberg, Germany. If you have any concerns regarding our products, please contact ProductSafety@springernature.com

Printed and bound by CPI Group (UK) Ltd, Croydon, CR0 4YY
01/10/2026
02225916-0007